American Institute of Chemical Engineers

Process Development Division

Presentations at the
2007 AIChE Annual Meeting

November 4-9, 2007
Salt Lake City, Utah, USA

Printed from e-media with permission by:

Curran Associates, Inc.
57 Morehouse Lane
Red Hook, NY 12571
www.proceedings.com

ISBN: 978-1-60560-009-3

Some format issues inherent in the e-media version may also appear in this print version.

ISBN: 978-1-60560-009-3

American Institute of Chemical Engineers
Proceedings
Three Park Avenue
New York, NY 10016-5991
Phone: 212-591-8100

www.aiche.org

American Institute of Chemical Engineers

Process Development Division

TABLE OF CONTENTS

Catalytic Partial and Complete Oxidation of Methane over the Catalysts with Ordered Macroporous Strucures in a Microchannel Chemical Reactor

Guoqing Guan, Katsuki Kusakabe* and Masatsugu Taneda*
Department of Living Environmental Science, Fukuoka Women`s University, 1-1-1 Kasumigaoka, Higashi-ku, Fukuoka 813-8529, Japan
guan@fwu.ac.jp or kusakabe@fwu.ac.jp

Abstract

The ordered macroporous ceramics were prepared on the wall of microchannel. The partial and complete oxidation reactions of methane over Rh-based and Pd-based catalysts supported on the ceramics were respectively investigated. Ceramic sol was infiltrated into the void of the polymethylmethacrylate (PMMA) opal structure formed in the microchannel. The dried sol in the opal structure was calcinated to fabricate the ordered macroporous supports. For comparison, the ceramic supports with random macropore were prepared by coating the suspension of PMMA particles, ceramic nanoparticles and binding materials on the wall of microchannel. The overall activity of the catalyst on the ordered macroporous support could improve due to the enhancement of heat and mass transfer. Rh-based and Pd-based catalysts supported on the prepared macroporous ceramics coated in the microchannels for the partial and complete oxidation of methane respectively were investigated.

Introduction

Microchannel chemical reactors show the promise for reducing the size of conventional chemical reactors. Partial oxidation of methane to syngas and the subsequent conversion of syngas to liquid fuels have been investigated extensively in connection with the exploitation of remote natural gas deposits. The process for the partial oxidation can benefit from using a microchannel chemical reactor due to its property of a fast exothermic reaction. Meanwhile, complete catalytic oxidation of methane in the microchannel could use for the energy supply to the micro fuel cell system.

Microreactors with microchannel ranging from 10 to 500 μm in diameter have proved to be suitable for fast reactions with a large heat effect because the small channel dimensions provide a relatively high surface area to volume ratio and increase the driving forces for heat and mass transport [1,2]. However, when a microreactor was used for the heterogeneous gas phase reactions, preparation and fabrication of solid catalysts in the microchannels were considered to be key issues. To date, wash-coating and dip-coating are generally utilized and a dense layer of catalysts can be successfully formed on the wall of microchannel by these methods [3,4]. Other improvement methods to increase the contact areas between the catalysts and reactants also appeared. For instance, Kenis *et al.* developed a method for preparation of SiCN and/or SiC catalyst supports with inverse opal structure in microchannels and used the obtained catalysts for the decomposition of ammonia and steam reforming of propane to produce hydrogen for fuel cells [5-7]. The catalyst with inverse opal structure can be used in practical process because of its high surface area per unit volume and low flow resistance. Recently, Al_2O_3 inverse opal was also fabricated in microchannels and Mo-promoted Pt catalyst supported on it had excellent reactivity for the propane combustions [8]. However, it is found that the prepared inverse opals in the microchannel did not strongly adhere to the wall of channel [5-8]. They had to be treated again by addition of ceramic glue to enhance the adhesive properties.

In the present study, we prepared macroporous ceramics in the microchannel by two methods; one is via infiltration of ceramic sol into the PMMA opal structures formed in the microchannel, and the other is via direct coating of the suspension composed of uniform template particles, ceramic nanoparticles and binding materials. The catalytic evaluation for the partial and complete oxidation of the methane was performed by using the precious metal catalysts with two kinds of structures and the results were compared with the catalysts without macroporous structures prepared by a wash coating method.

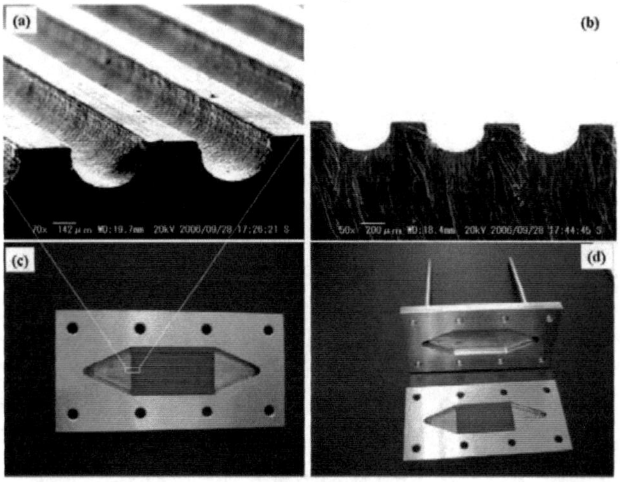

Fig.1 *(a) microchannel; (b)cross section of the channel ; (c) microchannel in jacket;(d) up and bottom jackets*

Experimental

Design of microchannel reactor The main body of microreactor was composed of two stainless steel plates (10 mm x 25 mm x 1 mm) with 14 microchannels (500 μm in width, 250 μm in depth and 25 mm in length) and put into stainless steel cells as shown in **Fig.1**. Before the catalyst coatings in the microchannels, the plates were cleaned and thermally treated at 800°C for 2h, followed by clean treatment and hydrophilic treatment.

Preparation of Al_2O_3 support with random macropores As shown in **Fig.2(a)**, 0.5 g of PMMA particles (Soken Chemical and Engineering Co., Japan) with a size of 310 nm were dispersed in 4 ml of distilled water containing 6 wt% of PVA (polyvinyl alcohol, Fluka). Then, 0.5 g of γ-Al_2O_3 nanoparticles were dispersed into it under stirring. After 24 h stirred at room temperature, the obtained suspension was introduced into the microchannels. It was found that the suspension can be moved along the microchannel automatically from one side to the other side. The coated plates were dried at 90°C for 12 h and then heated to 300°C in air and kept at this temperature for 2 h. The plates were finally calcinated at 600°C for 3 h.

Preparation of Al_2O_3 support with inverse opal structure As shown in **Fig.2(b)**, 10 wt% of PMMA colloidal suspension was introduced into the microchannel at first. After the water in the suspension was evaporated slowly at room temperature, PMMA opals with well-defined structure were formed in the microchannels. PMMA particle-packed microchannels were heated at 90°C for 24 h in order to improve the contact properties between the neighboring particles. Then, the prepared Al_2O_3 sol was infiltrated into the void in the PMMA opals. The infiltration process was repeated several times in a nitrogen filled glove box because of low concentration of Al_2O_3 sol. The heat treatment was the same as the above mentioned method.

Preparation of Al_2O_3 support without macro pores As shown in **Fig.2(c)**, γ-Al_2O_3 nano-particles were dispersed in a solution containing 6

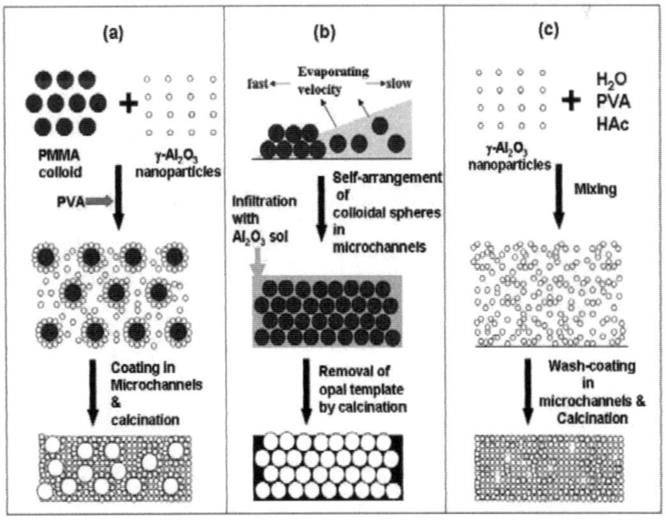

Fig.2 *Methods for the preparation of coatings in the microchannels (a) coatings with random macropores; (b) coatings with well-defined macropores; (c) coatings without macropores.*

wt% PVA and 2 wt% acetic acid under stirring. The obtained suspension was wash-coated in the microchannels according to the method reported in the literature [1-3]. The calcination method for the wash-coated plates was also the same as the above mentioned.

Impregnation of catalysts Incipient wetness impregnation of Rh or Pd catalyst on Al_2O_3 support in the microchannels for partial or complete oxidation of methane was performed as follows: the calculated amount of aqueous solution of $Rh(NO_3)_3$ or $Pd(NO_3)_2$ based on the weight of Al_2O_3 support was dropped into the coated microchannels, followed by drying at 120°C for 12 h, and then calcinated at 500°C for 3 h.

Characterization The morphologies of the prepared structure were observed with a scanning electron microscope (SEM, Hitachi, S-5200). Powder X-ray diffraction (XRD) patterns of the catalysts were recorded on a Rigaku RINT-2500 diffractometer using Cu Kαradiation (λ = 0.15418 nm) generated at 40 kV and 30 mA. Nitrogen absorption/desorption isotherm measurements were done at 77K on a Micrometritics ASAP 2010 gas adsorption analyzer. Prior to the measurements, the samples were degassed at 120°C for 12 h.

Catalytic reaction For partial oxidation of methane, the feed gas containing 29.4% CH_4, 14.7% O_2,and 55.9% N_2 was passed through the microreactor with a total flow rate of 50 ml·min^{-1} (STP). For complete oxidation of methane, the feed gas containing 9.5% CH_4, 19% O_2, and 71.5% N_2 was passed through the microreactor with a total flow rate of 90 ml·min^{-1} (STP). H_2, CO, CO_2 and CH_4 in the effluent gases were analyzed by a gas chromatograph (GC-8A, Shimadzu) equipped with a thermal conductivity detector (TCD). In order to detect low concentration levels of CO and CO_2, a methanizer (MT221, GL Sciences) packed with a nickel catalyst powder was installed on the gas line before a flame ionization detector.

Results and discussion

Figs.3 (a)-(d) show the SEM images of γ-Al_2O_3 coatings in the microchannels using the three methods. When the suspension containing PMMA particles, Al_2O_3 nanoparticles and PVA was used for the coating, random macropores were formed in the coating layer as shown in **Fig.3 (a)**. Macropores distributed uniformly, and three dimensional porous structures were partially observed. The existence of these macropores in the coating layer was expected to improve the mass transfer between the gas phase and the catalyst surface, enhancing the reactivity in gas phase reactions. In addition, the random macroporous structure indicated the strong adhesion to the wall of microchannels. **Fig.3 (b)** shows the image of PMMA opals formed in the microchannel, as a template. When the prepared Al_2O_3 sol was infiltrated into the voids of PMMA opals and calcinated, well-defined ordered macropore structures (inverse opals) of Al_2O_3 were formed as shown in **Fig.3(c)**. Because of this well-ordered structure, the mass transfer could be further improved. For comparison, the image of wash-coating Al_2O_3 layer in the microchannel is shown in **Fig.3(d)**. In this case, no any special macropore was formed in the coating layer.

The catalytic partial oxidation of methane ($CH_4+0.5O_2=CO+2H_2$) can be considered to be composed of the following reaction steps: in step 1: $CH_4+2O_2=CO_2+H_2O$, i.e. methane oxidation step; in step 2, two reactions could be taken into account: (1) $CH_4+CO_2=2CO+2H_2$, dry reforming reaction and (2) $CH_4+H_2O=CO+3H_2$, steam reforming reaction, and herein, we call two reactions as reforming reactions step. Partial oxidation of methane over the catalyst of 5wt% Rh/Al_2O_3 with a random macropore structure was investigated at first. As shown in **Fig.4**, the methane conversion rate as well as the H_2 and CO selectivities increased with increase in the reaction temperature. This means that the reforming reactions with high activation energies compared to methane oxidation proceeds preferentially at high temperature. Furthermore, as indicated in the inset figure of **Fig.4**, when the reaction temperature increased over 800 K, the H_2/CO molar ratio in the product became approximately 2.0, which was close to the optimal ratio for low-temperature Fischer-Tropsch reactions to paraffinic oil. According to the reported equilibrium values determined by thermodynamic calculation method for this reaction, the CH_4 conversion and the H_2 and CO selectivities should reach 100% at 1273K, however, in the present study, the thermodynamic

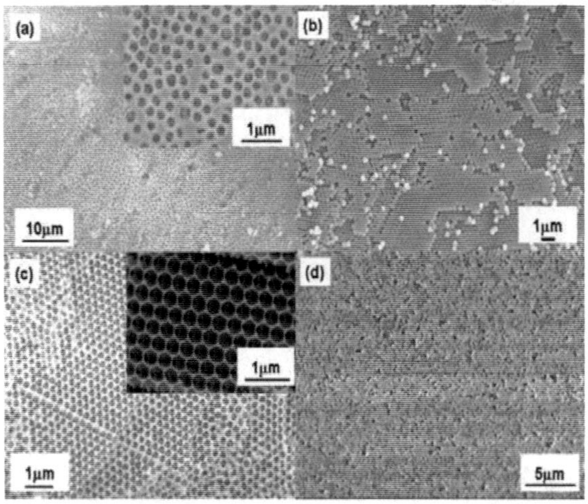

Fig.3 *SEM images of (a) Al_2O_3 coating with random macropore structure; (b) PMMA opals formed in the microchannel;(c) Al_2O_3 coating with ordered macropore structure; (d) Al_2O_3 coating without macropores.*

equilibrium was not achieved. Fichtner *et al.* fabricated microstructured honeycomb catalyst of Rh foils and used the catalyst for the partial oxidation of methane [9].They found that hot spots were prevented in the Rh honeycombs because of their high axial thermal conductivity. Meanwhile, methane conversion and hydrogen selectivity were much higher than the published data obtained with ceramic honeycombs. Tonkovich *et al.* packed Rh/SiO_2 catalyst powders in the microchannel reactors and used for the partial oxidation of methane [10]. According to their results, thermodynamic equilibrium was also not achieved. However, the present data of CH_4 conversion as well as the H_2 and CO selectivities was higher than that reported in literatures [9,10] at the same reaction temperature, suggesting the increase in the catalytic activity due to the macropores introduced in the catalyst structures. After the reaction temperature was raised to 1273K, the reactor was damaged. Therefore, the reactions over 5 wt% Rh/Al_2O_3 catalysts with ordered macropore structure and without macropore structure were performed below 923K. **Fig.5** reveals that the CH_4 conversion for the catalyst with ordered macropore structure was slightly high due to the high porosity compared to the catalyst with random macropore structure. When the catalyst without macropore structure was used in the tested temperature range, the CH_4 conversion was lowest. Accordingly the catalytic activity could be improved by introduction of macropore in the catalyst support on the microchannel.

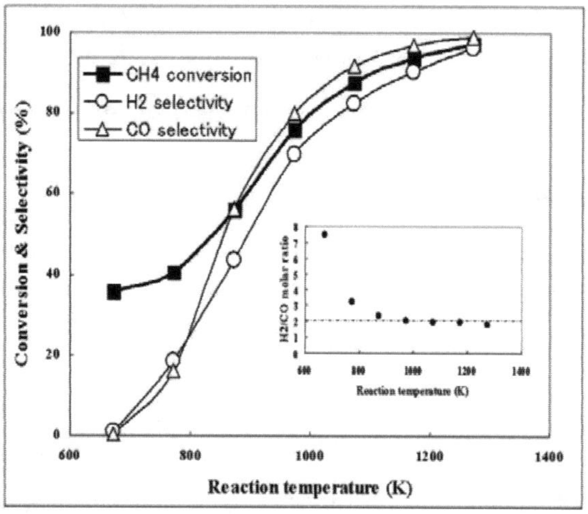

Fig.4 *CH_4 conversion rates and H_2, CO selectivities of partial oxidation of methane at CH_4/O_2 ratio of 2 as a function of the reaction temperature over 5wt%Rh/Al_2O_3 catalyst with random macropore structure. Inset shows H_2/CO molar ratio in the product as a function of the reaction temperature.*

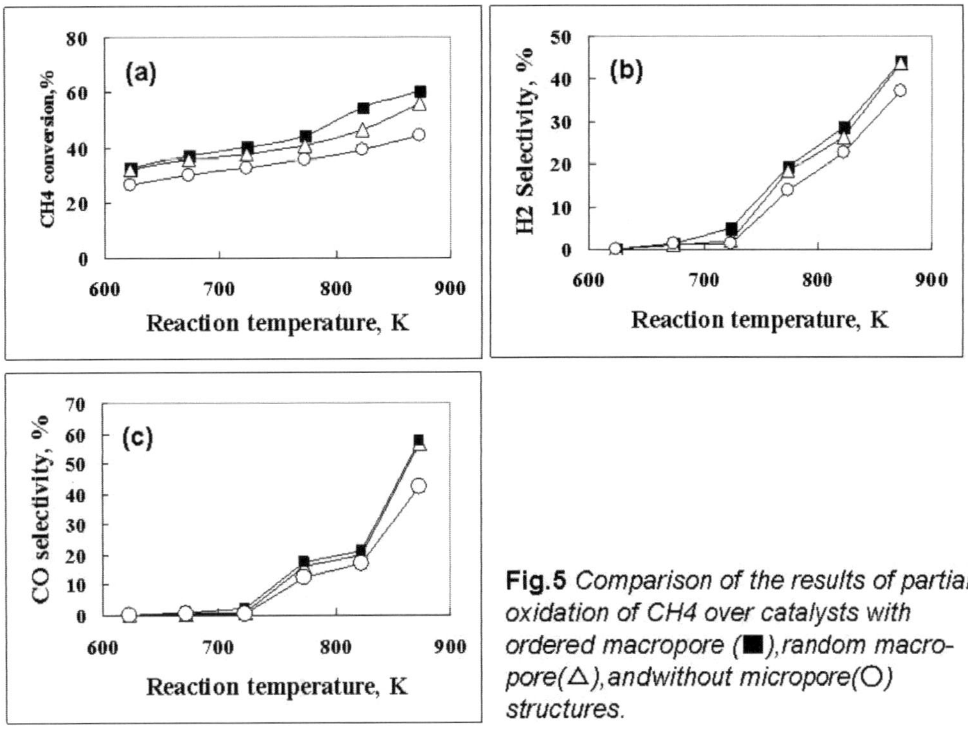

Fig.5 *Comparison of the results of partial oxidation of CH4 over catalysts with ordered macropore (■),random macro-pore(△),andwithout micropore(○) structures.*

Fig.6 shows complete oxidation of methane over 5wt%Pd/Al_2O_3 catalysts with ordered, random macropore structure and without macropore structure as a function of reaction temperature. As a result of the introduction of macropore in the catalyst support, the reactivity was highly enhanced.

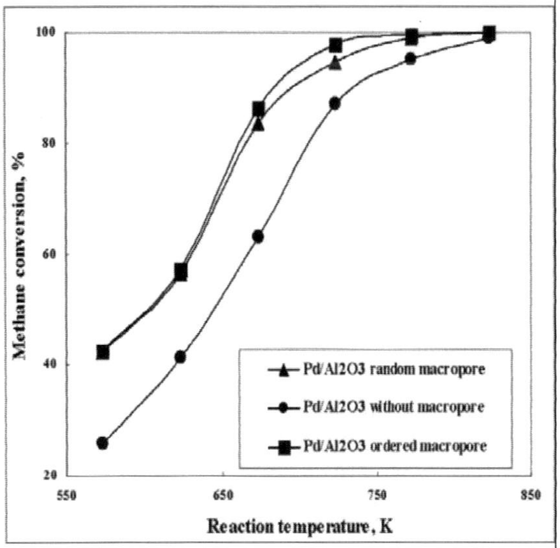

Fig.6 *Complete oxidation of methane over the catalysts with various macroporous structures.*

Conclusion

Al_2O_3 catalyst supports with ordered and random macropore structures and without macropore structure were fabricated in the microchannels. Partial oxidation of methane over the Rh/Al_2O_3 catalysts and complete oxidation of methane over the Pd/Al_2O_3 catalysts were investigated. The catalyst support with inverse opal structure had high porosity and highly uniform morphology in the microchannel. Therefore the precious metal catalysts highly dispersed on the supports with ordered macropore structure showed highest activity for the partial and complete oxidations because of the enhancement of gas-solid contact.

Acknowledgements

This study was supported by Japan Society for the Promotion of Science (JSPS).

Reference cited

1. V. Hessel, H. Löwe, A. Müller, G. Kolb, *Chemical Micro Process Engineering – Processing,* Applications, and Plants, Wiley-VCH, Weinheim, Germany, **2005.**
2. G. Kolb, V. Hessel, *Chem. Eng. J.* 98(2)(**2004**)1.
3. R. Zapf, et al., *Trans IchemE,* 81(A)(**2003**)721.
4. K. Haas-Santo, M. Fichtner, K. Schubert, *Appl. Catal. A,* 220(**2001**)79.
5. I.-K. Sung, et al., *Adv. Funct. Mater.* 15(**2005**)1336.
6. Christian, M. Mitchell, D.-P. Kim, P. J. A. Kenis, *J. Catal.* 241(**2006**)235.
7. Christian, M. Mitchell, P. J. A. Kenis, *Lab Chip,* 6(**2006**)1328.
8. G. Guan, R.Zapf, G. Kolb, et al., *Chem. Commun.***2007**, 260.
9. M Fichtner et al. *Ind. Eng. Chem. Res.* 40**(2001**)3475
10. A. Y. Tonkovich, et al. The Catalytic Partial Oxidation of Methane in a Microchannel Chemical Reactor, Published in the *Proceedings of the Second International Conference of Microreaction Technology,* March **1998**, New Orleans, Louisiana.

Production Of Zirconia Nanoparticles Using A Microreactor With Same Axle Dual Pipe

Taisuke Maki, Junichi Kitada, Kazuhiro Mae, Kyoto University, Japan

Introduction

Nanoparticles are receiving interesting attention because of their distinct physical and chemical properties. It is expected that nanoparticles will be applied to many new and high-functional devices. Considering such an application of the nanoparticles, it is important to develop a manufacturing process for the nanoparticles with narrow size distribution. In recent years, fine particles, which diameters are 1 μm or less, and monodispersed particles have been developed to advance their functionalities and add value to chemical industry products. In particular, monodispersed ultrafine particles, which diameters are 100 nm or less, are highly required. The precise control of the particle properties is, therefore, an extremely important technology for industrial use. However, it is difficult to control the reaction precisely in a conventional macroreactor because of the non-uniformity of the residence time, mixing and so on. Generally, it is essential to design precise nuclei formation and aggregation processes for high quality particles. The way to form the nuclei is to keep a proper super-saturation state by controlling the gradient of temperature, concentration of reactant and so on. From this viewpoint, a microchannel is very attractive for easily forming the reaction conditions in continuous production. Many researchers on fine-particle formation have been performed with microreactors in recent years [1-5]. Schenk *et al.* [1] developed two types of the separation layer mixer, named concentric and stacked separation layer micromixer, and presented a new concept for antifouling microchemical processing with solids precipitation from experimental and numerical approaches. There are two types of mixing in microreactors: convection mixing and precise mixing via interface by controlled-molecular diffusion. Many experimental and numerical researches have been performed on fine particle formation in convection mixing. For example, Schwarzer *et al.* investigated the relation between the mixing and the particle size and distribution in fine particles formation [6, 7]. They conducted fine particles formation of barium sulfate utilizing a microreactor with T-shaped channel, and estimated the particle size and distribution of the fine particles. On the other hand, there have been several researches on fine-particle formation in precise mixing via interface by controlled molecular diffusion. Wille *et al.* used the multilamination type microreactor developed by CPC-Systems to form ultrafine pigment particles of a mean-particle size of 90 nm, and showed that the particles are smaller in size than those formed by the conventional batch method [8].

Mae *et al.* [9] and Nagasawa *et al.* [10] have also developed some micromixers for convection and concentric mixing. We conducted fine particles formation of polystyrene by using two types of micromixers for convection mixing, which were named YM mixer and KM mixer that utilizes splitting and recombination of fluids and microsegments collision, respectively, and showed that both mixers were useful as a reactor for forming fine particles with narrow distribution. Moreover, we have presented a new microreactor design concept for particle production. The idea lies in a realization of microspace by an annular laminar flow of two immiscible liquids to prevent the particle precipitation on the wall and to control particle formation. The proposed device was confirmed to be valid for the mono-modal and spherical particle formation, with the result that quite uniform spherical particles of titania were produced without precipitation of particles at the wall. We successfully controlled the mono-modal particles from 5 to 150nm by controlling the residence time to the sub-second and other reaction condition. Also, it was shown that this microreactor was effective to estimate the nucleation region, which is an important factor in particle formation, the particle-formation -frequency distribution in the region, and the behavior of the particle growing process [11-13]. In this study, we extended to the above microreaction technology to nano-particle formation with more rapid reaction rate. As a representative example, zirconia particles were produced by supplying zirconium tetrabutoxide solutes to the inner tube and alcohol/water mixture to the outer tube. We examined the effect of solvent and alcohol type, the micro channel size and the residence time on particle properties.

Experimental

Experimental setup

Figure 1 shows the schematic drawing of concentric microreactor used. The microreactor consists of a transparent glass pipe and a stainless pipe. The pipes are set coaxially so as to form a dual pipe structure. The length (*L*) of the microchannel can be changed from 0.5 to 2.0 cm. Two reactants were fed in the inner and outer tubes, respectively, and kept an annular and laminar flow of separated phase. In the flow, the inner flow corresponded to a microchannel, which could be changed by operation conditions. On the basis of this concept, the reaction can be precisely controlled by the gradient of temperature and/or concentration of reactant in the outer fluid. In addition, the outer fluid plays a role of preventing precipitated particles from adhering to the tube wall. The device also has several advantages such as easy maintenance, cheap setting, etc.

Preparation of zirconia particles

In this study, we conducted zirconia fine particle formation as an example having very rapid reaction rate. The reaction formulas are given in Eqs. 1 and 2.

$$\mathrm{Zr(OC_4H_9)_4 + 4H_2O \rightarrow Zr(OH)_4 + 4C_4H_9OH} \quad (1)$$

$$\mathrm{Zr(OH)_4 \rightarrow ZrO_2 \cdot xH_2O + (2-x)H_2O} \quad (2)$$

Alkoxide hydrolysis has merits of high productivity, high purity production, and so on for industrial use. However, its high reaction rate makes it difficult to control particle size and shape. To reduce the reaction rate, alcohol solution (Reactant A) of zirconium tetrabutoxide (ZTB: $Zr(OC_4H_9)_4$) and alcohol aqueous solution (Reactant B). Reactant A and B were supplied to the inner tube and to the outer tube of the microreactor, respectively, with syringe pumps. **Figure 2** shows a presumed mechanism for the nucleation and the growth of particles in the microreactor in our previous work [12]. The nucleation and the particle growth proceed on the interface of two fluids. In the nucleation zone, nuclei are generated with H_2O diffused to the interface between fluids at an initial zone of the microreactor. In the particle growth zone, on the other hand, particles were grown uniformly by the ordered flux of alkoxide into the interface.

To show the superiority of the fine-particle formation technique with the microreactor, we also conducted an experiment by using a conventional method with a typical batch reactor for comparison.

Thus prepared particles were observed with a transmission electron microscope (TEM, JEOL JEM-1010). The mean size and distribution of formed zirconia particles were analyzed by a dynamic light scattering nanoparticle size analyzer (DLS, HORIBA LB-550).

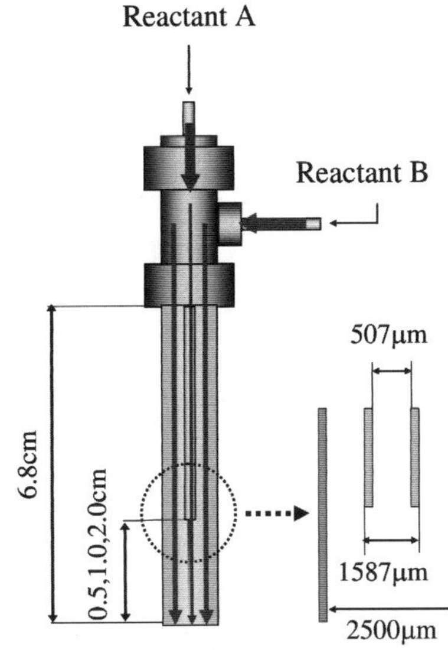

Figure 1 Schematic of the microreactor.

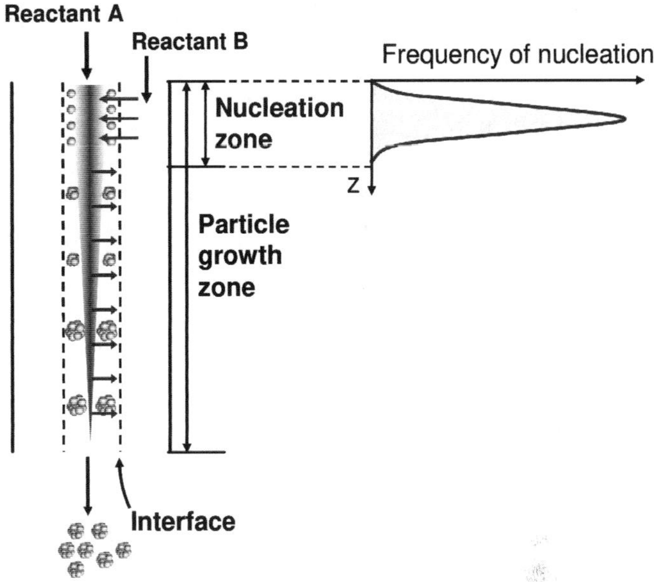

Figure 2 Presumed mechanism on fine-particles formation in the microreactor.

Results and Discussion

Production of zirconia fine particles using the concentric microreactor

We first examined the conditions (alcohol type, concentration of ZTB and water *etc.*) for production of fine zirconia particles in a conventional-stirred vessel with a batch method under a sufficiently strong stirring condition. Only when ethanol was used as a solvent of zirconium tetrabutoxide, spherical zirconia particles were produced. **Table 1** shows the effect of alcohol type for Reactant B (aqueous solution) on the particle size distribution.

Table 1 Effect of alcohol type for Reactant B on the particle size distribution.

Alcohol type	Median size [nm]	CV [%]
Ethanol	79.2	34.5
1-Propanol	81.0	27.7
1-Buthanol	116.9	28.9
1-Hexanol	146.6	24.7
1-Octanol	110.9	174.2

As carbon number of alcohol used increased, the mean size of particles increased. The particle size distribution for 1-octanol was significantly broad. These results indicate that produced zirconia fine particles are stable in lower alcohol. Generally, the viscosities of straight-chain alcohol increase with increase in carbon number. For a highly viscous solvent such as 1-octanol, the flux of ZTB decreased by low diffusion rate and the aggregation between primary particles was dominant, leading to random particles.

Next, we focus on the effect of each reactant concentration. In **Figure 3**, the median diameter increased monotonically as the ZTB concentration increased. In the particle-growth process, the higher the ZTB concentration was, the larger the concentration gradient was, and hence, the larger the diffusion rate of ZTB was around the nuclei. This means that the amount of supplied ZTB per unit number of nuclei was large. As a result, nucleus growth was promoted. On the other hand, the particle size was smallest when the H_2O concentration of Reactant B was around 20 wt%. Similar results were obtained in the case of titania particles formation [12]. When the H_2O concentration was smaller than 20 vol.%, the degree of supersaturation in the nucleation process reduced due to a lack of H_2O supply, and a small number of nuclei are formed. On the contrary, when the H_2O concentration was higher than 30 vol.%, the degree of supersaturation in the nucleation process increased due to an excessive H_2O supply, and many nuclei were formed. However, in the particle growth process of these nuclei, excessively supplied H_2O enhanced the dielectric constant around the nuclei, and the nucleus aggregation occurred at random.

From the above examination, we determined that Reactant A and B were ethanol solutions including 0.05 M of ZTB and 3 wt.% of H_2O, respectively, for the production of zirconia particles by using the concentric microreactor. **Figure 4** is a TEM photograph of zirconia particles produced by the microreactor. Zirconia particles were produced and grown with spherical shape in an inner tube and no precipitation of particles was observed at the wall by the outer fluid. **Figure 5** compares the size distributions of the fine particles formed by both the microreactor and the batch reactor. The size distribution of fine particles formed by the microreactor was narrower than that formed by the stirred vessel. A similar tendency to our other findings were obtained [11, 12], indicating that the concentric microreactor was effective for very rapid reaction as hydrolysis of zirconium alkoxide. In this system, since the mixing of the reactants proceeds mainly by the molecular diffusion via interface in the laminar flows and the process of the nucleation and particle growth in the fine-particle formation was more ordered than that in the random mixing by batch method.

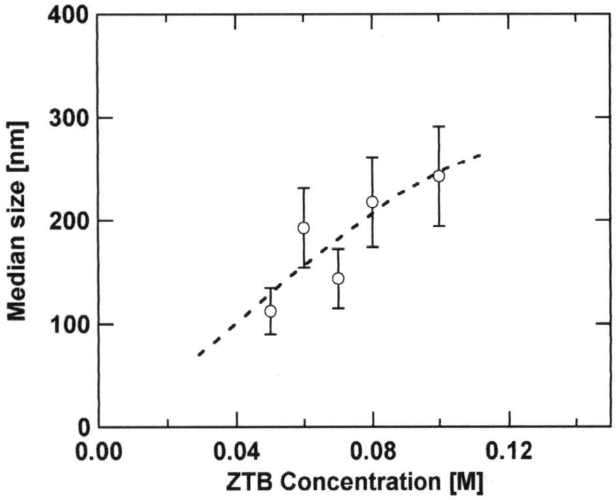

Figure 3 Influence of the ZTB concentration.

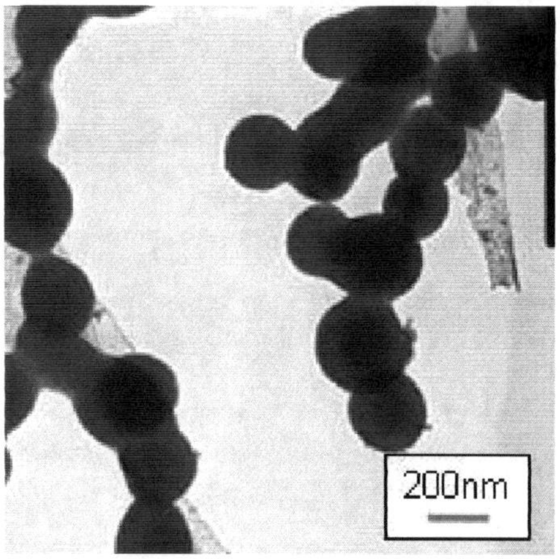

Figure 4 TEM photograph of zirconia fine particles produced by the microreactor.

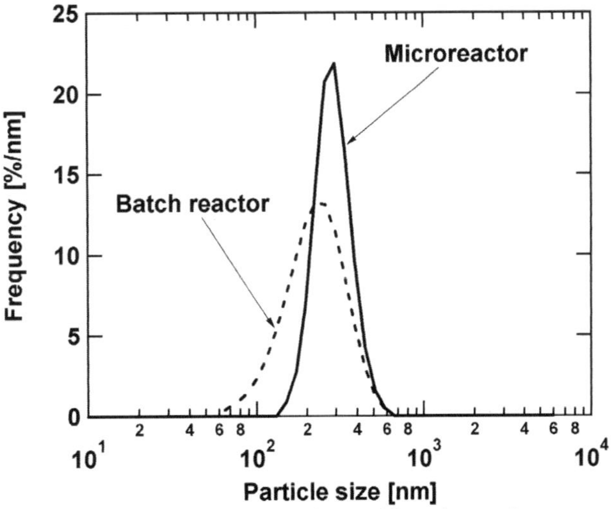

Figure 5 Comparison of the fine particles produced by the microreactor and the batch reactor.

Effect of the operation parameters

To clarify that the concentric microreactor had the possibility of controlling the growth of zirconia nanoparticles, we examined the effect of the residence time determined by following two parameters; a length of reaction zone, L, and a mean linear velocity of fluids, $V_{ave.}$. **Figure 6** shows the change in the median diameter and the coefficient of variation, CV, with the residence time. From Figure 6a, the median diameter increased monotonously with an increase in L, namely, the residence time. This may be because the nuclei formed near the reaction interface were grown with the residence time by supplying the reactants through molecular diffusion. We conducted the experiments with several residence times by changing mean linear velocity, and found that the median diameter had minimum value at $V_{ave.}$ = 1.0 m/s as shown in Figure 6b. This may be because ZTB was not perfectly reacted when $V_{ave.}$ is more than 1.0 m/s. On the other hand, the CV values were almost constant, irrespective of the residence time except for an unstable flow region at L= 0.5cm. The fact indicates that the

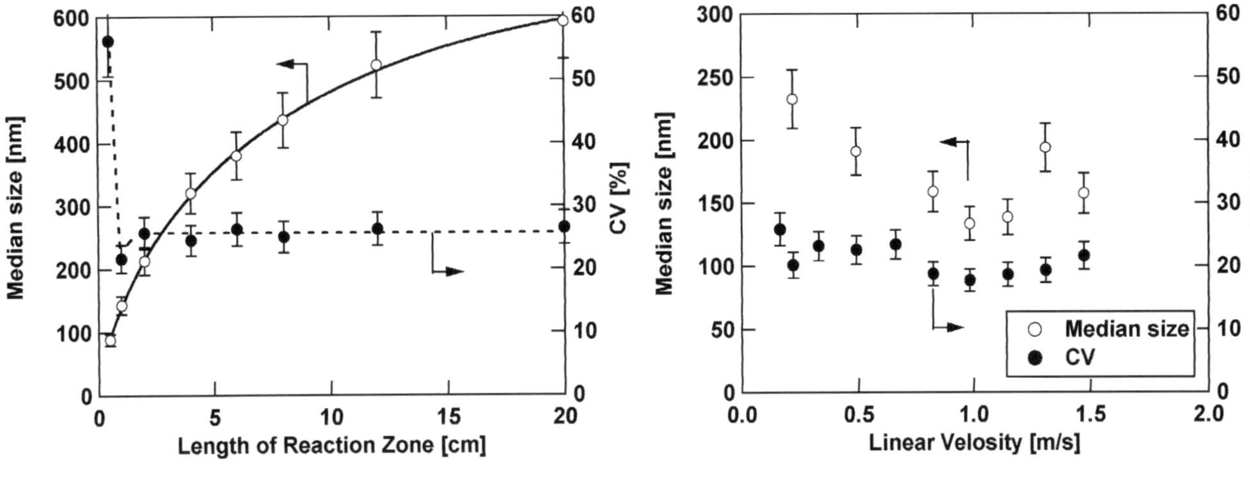

Figure 6 Change in the median diameter and the coefficient of variation with the residence time:

nuclei formation and particle growth can be separately controlled, namely, the number of nuclei having uniform size and the final particle size are strictly determined by the flux in the nucleation zone and the particle growth zone, respectively.

Figure 7 shows the change of median size at very short residence time. Even when the residence time was more than 10ms, the median diameter of zirconia particles increased monotonously. The result indicates that the proposed microreactor has a potential to precisely control very fast reactions within 100 ms of reaction time by controlling molecular diffusion.

Possibility of nanoparticle formation

Finally, we tried to form nanoparticles using this concentric microreactor. In our previous work [12], we successfully produced the titania nanoparticles of 3 nm with a narrow distribution by reducing the diffusion rate of titanium tetraisopropoxide (TTIP) under the low TTIP concentration in the concentric microreactor. However, the nanoparticles of zirconia could not be produced only by changing reaction conditions because of its extremely fast reaction rate. Thus, several types of dispersants were added into Reactant B to suppress the particle growth, and it was found that polyethyleneimine (PEI) was very effective for the stabilization of zirconia particles. **Figure 8** shows the effect of the concentration of PEI on the median diameter of zirconia particles. The median diameter decreased with an increase in the PEI concentration. The minimum size of 4nm was obtained at a high PEI concentration of 0.24 wt.%. Thus, we successfully controlled the mean particle size of zirconia in the range from 4nm to 700nm simply by changing the mean residence time and the concentrations of zirconium tetrabutoxide and H_2O in the presence of a dispersant, polyethyleneimine.

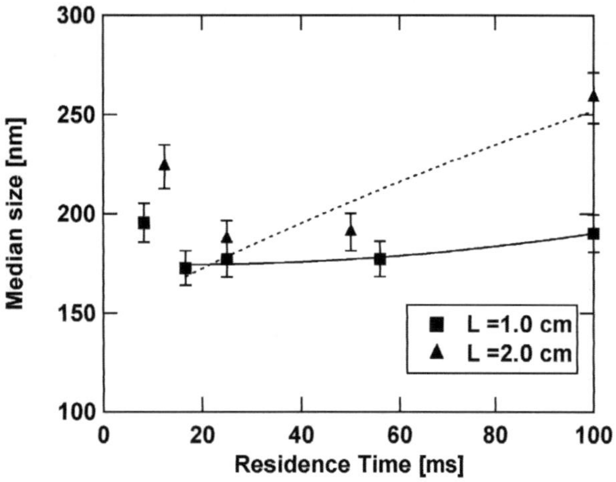

Figure 7 Median size of zirconia particles under short residence time

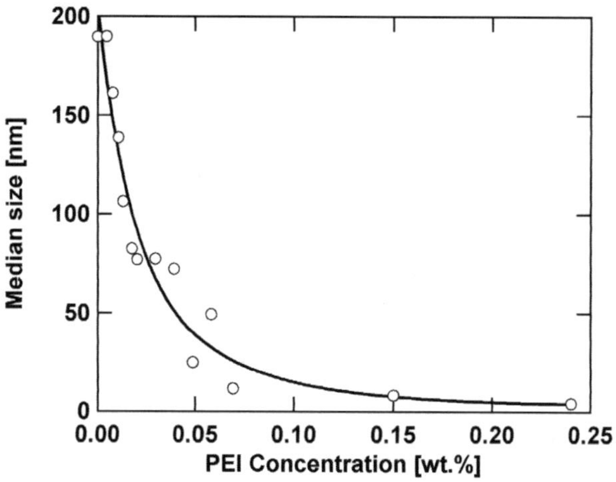

Figure 8 Effect of PEI concentration on on the median diameter

Conclusion

We conducted a zirconia fine particle formation experiment using a concentric microreactor with same axle dual pipe developed in our previous work. Zirconia particles were produced and grown in an inner fluid and no precipitation of particles was observed at the wall by sealing of an outer fluid. The size and shape of zirconia particles produced was significantly differed by the solvent and alcohol type. Only when ethanol was used as a solvent of zirconium tetrabutoxide, spherical zirconia particles were produced. As carbon number of alcohol used increased, the mean size of particles increased. When ethanol was used as a solvent for diluting water, the zirconia particles had a sharp and narrow distribution which could not be realized by conventional batch method. In addition, the mean particle size was successfully controlled in the range from 4nm to 700nm simply by changing the mean residence time and the concentrations of zirconium tetrabutoxide and water in the presence of a dispersant, polyethyleneimine. It is suggested that the primary nucleation and particle growing can be independently and precisely controlled for the very fast reaction within 100 ms of reaction time, through adjusting the fluxes of reactants under laminar flow in the concentric microreactor.

References

1. Schenk R., Hessel V., Werner B., Schönfeld F., Hofmann C., Donnet M., Jongen N. (2002), "Micromixers as a tool for powder production," *Chem. Eng. Trans.*, 1, pp. 909–914.
2. Nakamura H., Yamaguchi Y., Miyazaki M., Maeda H., Uehara M., Mulvaney P. (2002), "Preparation of CdSe nanocrystals in a micro-flow-reactor," *Chem. Commun.*, 23, pp. 2844–2845.
3. Wang H., Li X., Uehara M., Yamaguchi Y., Nakamura H., Miyazaki M., Shimizu H., Maeda H. (2004), "Continuous synthesis of CdSe-ZnS composite nanoparticles in microfluidic reactor," *Chem. Commun.*, 1, pp. 48–49.
4. Mae K., Maki T., Hasegawa I., Eto U., Mizutani Y., Honda N. (2004), "Development of a new micromixer based on split/recombination for mass production and its application to soap free emulsifier," *Chem. Eng. J.*, 101, pp. 31–38.
5. Wille Ch., Gabski, H-P., Haller Th., Kim H., Unverdorben L., Winter R. (2004), "Synthesis of pigments in a three-stage microreactor pilot plant -an experimental technical report," *Chem Eng. J.*, 101, pp. 179–185.
6. Schwarzer H-C., Peukert W. (2004), "Combined experimental/numerical study on the precipitation of nanoparticles," *AIChE J.*, 50, pp. 3234–3247.
7. Schwarzer H-C., Schwertfirm F., Manhart M., Schmid H-J., Peukert W. (2006), "Predictive simulation of nano particle precipitation based on the population balance equation," *Chem Eng Sci.*, 61, pp. 167–181.
8. Wille Ch., Autze V., Kim H., Nickel U., Oberbeck S., Schwalbe Th., Unverdorben L. (2002), "Progress in Transferring Microreactors from Lab into Production an Example in the Field of Pigments Technology," *Proceedings of the 6th International Conference on Microreaction Technology*, New Orleans, LA., pp. 7–17.
9. Mae, K., Maki, T., Hasegawa, I., Eto, U., Mizutani, Y.. Honda, N.(2004), "Development of a New Micromixer Based on Split/Recombination for Mass Production and its Application to Soap Free Emulsifier", *Chem.l Eng. J.* **101**, 31–38.
10. Nagasawa H., Aoki N., Mae K. (2005), "Design of a new micromixer for instant mixing based on the collision of micro segments," *Chem. Eng. Technol.*, 28, pp.324–330.
11. Takagi M., Maki T., Miyahara M., Mae K. (2004), "Production of titania nanoparticles by using a new microreactor assembled with same axle dual pipe," *Chem. Eng. J.*, 101, pp. 269–276.

12. Nagasawa H., Tujiuchi T., Maki T., Mae K. (2007), "Controlling Fine Particle Formation Processes using a Concentric Microreactor," *AIChE Journal*, **53**, 196–206.
13. Nagasawa, N., Mae, K. (2006),"Development of a New Microreactor Based on Annular Microsegments for Fine Particle Production ", *I & EC Res.*, **46**, 2179–2186.

64c A Mems-Reformer-Based Integrated System for on-Site Power Generation from Ammonia Decomposition

Shihuai Zhao, Sagar Gururaj, Naveed Ansari, William R. Ashurst, and Bruce Tatarchuk. Chemical Engineering, Auburn University, Auburn, AL 36849

Hydrogen has been recognized as the most attractive alternative and renewable energy due to consumption of gasoline and environmental concern. Ammonia decomposition provides a direct and simple route for hydrogen production since no COx and H2S are involved, which means that no water-gas shift reactors, preferential oxidation reactors and sulfur removal units are required. In this study, a MEMS reformer based on Si as a substrate material has been developed for hydrogen production from ammonia decomposition. Silica supported Pt or Ru was chosen as reforming catalyst in form of thin film and microfibrous entrapped particles, and prepared by incipient wetness impregnation. The reaction of ammonia decomposition was tested in temperature range of 200 – 600 °C. The microfibrous entrapped catalyst produced higher conversion of NH3 at all the temperatures compared to thin film catalyst, and Ru is better catalyst than Pt. For microfibrous entrapped catalysts, the experiments on different sizes of silica particles show that 20 μm silica produced higher NH3 decomposition than 40 and 60 μm particles. Ammonia conversion on microfibrous entrapped Ru reached 90% at 500 °C, and was close to 100% at 600 °C.

64d Catalyst Aided Nanomaterial Growth In Microreactor Chamber Under Extreme Conditions

Silpa Kona[1], Cindy Harnett[1], and Mahendra K. Sunkara[2]. (1) Department of Electrical and Computer Engineering, University of Louisville, 106 Ernst Hall, Department of Chemical Engineering,, University of Louisville, Louisville, KY 40292, (2) Department of Chemical Engineering, University of Louisville, 106 Ernst Hall, Department of Chemical Engineering,, University of Louisville, Louisville, KY 40292

Microreactors offer limitless possibilities when compared to their macroscale counterparts. The small volumes involved in these microchamber reactors facilitate reactions to take place under extreme conditions like very high pressures and temperatures. Here, we present nanomaterial growth under these conditions in the presence of a catalyst. The substrate being the microreactor chamber allows flexibility in terms of chamber volume and gives scope to experiment with different precursor gases

Enhanced Performance of PID Controller Design for Second–Order Processes with Time Delay

M. Shamsuzzoha[1], *Moonyong Lee*[2]*

School of Chemical Engineering and Technology, Yeungnam University, Kyongsan, 712-749, Korea, [1]*(E-mail: smzoha2002@hotmail.com),* *[2]*(mynlee@yu.ac.kr)*

Abstract: A design method for a proportional–integral–derivative (PID) cascaded with a lead-lag compensator is proposed for enhanced disturbance rejection of the second-order stable and unstable processes with time delay. A two-degree-of-freedom control scheme is used to cope with both regulatory and servo problems. An ideal feedback controller equivalent to IMC is obtained through the IMC design principle, and is further simplified to the PID cascaded with a first-order lead-lag compensator. The simulation is conducted for a broad class of stable and unstable processes and the results are compared with those of recently published PID type controllers to illustrate the superiority of the proposed controller. For a reasonable comparison, the controllers in the simulation study are tuned to have the same degree of robustness by measuring the maximum sensitivity, Ms . The robustness of the controller is also investigated by simultaneously inserting a perturbation uncertainty in all parameters in order to obtain the worst-case model mismatch. The proposed method illustrates greater robustness against process parameter uncertainty.

Introduction

It is common practice to reduce the actual process to a low-order model because most tuning rules for the proportional–integral–derivative (PID) type industrial controller are based on such low-order models. It is also well known that the second–order plus time delay model represents the dynamics of the actual process better than the first–order plus time delay model for various processes in chemical process industries. Furthermore, in chemical process industries, the majority of control loops is of the PID type due to its relatively simple structure, which can be easily understood and implemented in practice.

To find a simple design method of the PID type controller with a significant performance improvement has become an important research issue for process engineers. Due to the simplicity and improved performance of the internal model control (IMC)-based tuning rule, the analytically derived IMC–PID tuning methods have attracted the attention of industrial users over the last decade. The IMC–PID tuning rule has only one user–defined tuning parameter, which is directly related to the closed–loop time constant. The IMC–PID tuning methods[1-4] and the direct synthesis (DS) method[5] are two examples of typical tuning methods based on achieving a desired closed–loop response.

The IMC–PID controller provides good set–point tracking but sluggish disturbance response, especially for processes with a small time–delay/time–constant ratio. However, for many process control applications, disturbance rejection is much more important than set–point tracking. Several researchers[1-5] have reported that the suppressing load disturbance is poor when the process dynamics are significantly slower than the desired closed–loop dynamics. Therefore, a controller design emphasizing disturbance rejection rather than set–point tracking is an important design problem that has received renewed interest recently. Furthermore, it is well known that the control system design for an open–loop, unstable process is even more difficult especially in the presence of a time delay. Therefore, some modified IMC methods[6-10] of two–degree–of–freedom (2DOF) control were developed for controlling unstable processes with time delay. Recently, Rao and Chidambaram[11] proposed a PID controller in series with a lead-lag compensator for the open–loop unstable, second-order plus time delay processes with/without a zero.

The PID controller design has been discussed extensively in the literature for both stable and unstable processes, but the design of a simple and robust controller with improved performance has not yet been fully achieved. The existing modern control hardware provides microprocessor implementation for a flexible combination of conventional control algorithms to achieve enhanced control performance. The PID controller in series with a lead–lag compensator (hereafter, PIDC controller) is a typical example. The main reason for using the PIDC controller is to provide improved performance without tribulation. Many authors[1-2,6,11-12] have proposed this type of PIDC control strategy, as described in Eq. (1) below.

$$G_c = K_c\left(1+\frac{1}{\tau_I s}+\tau_D s\right)\frac{1+as}{1+bs} \tag{1}$$

where K_c, τ_I and τ_D are the proportional gain, integral time constant, and derivative time constant of the PID controller, respectively, and a and b are the compensator parameters.

The present study focused on the design of the general second-order process in Eq. (2) to fulfill the various objectives: the tuning rule should be simple, analytically derived, model–based, and able to be applied to several process categories with a wide range of variation in process parameters.

$$G_p = G_D = \frac{(\tau_a s \pm 1)Ke^{-\theta s}}{(\tau_1 s \pm 1)(\tau_2 s \pm 1)} \tag{2}$$

The proposed method is extended for the first-order delay integrating processes with dead time. The set–point filter is used to eliminate the overshoot in set–point response. The performance of the proposed tuning rule has been compared with other prominent tuning methods in such a way that each controller is tuned to the same robustness level by evaluating the peak of the maximum sensitivity (M_s).

PIDC Controller Design Method

The closed–loop control block diagram is shown in Fig. 1, where G_p is the process transfer function, $\tilde{G}_p$ the process model, q the IMC controller, f_R the set–point filter, and G_c the transfer function of the controller.
For the nominal case (*i.e.*, $G_p = \tilde{G}_p$), the set–point and disturbance responses in the IMC control structure can be simplified as:

$$y = G_p q f_R r + (1 - \tilde{G}_p q) G_D d \tag{3}$$

In accordance with the IMC method[1], the process model $\tilde{G}_p$ is factored into two parts:

$$\tilde{G}_p = p_m p_A \tag{4}$$

where p_m and p_A are the portions of the model inverted and not inverted by the controller, respectively, and $p_A(0) = 1$. The noninvertible part usually includes the dead time and/or right half plane zeros and is chosen to be all–pass.
To obtain a good response for processes with negative poles or poles near zero, the IMC controller q should be designed to satisfy the following conditions.

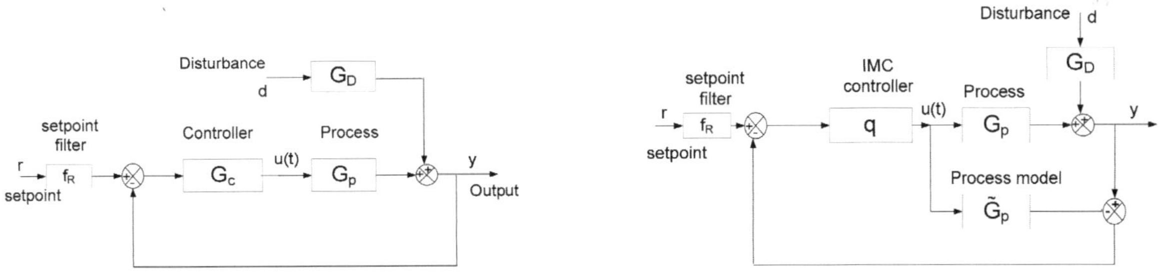

(a). Classical feedback control structure (b). The IMC structure

Figure 1. Block diagram of IMC and classical feedback control.

Unstable/Stable Processes with Poles near Zero

To get a superior response for unstable processes or stable processes with poles near zero, the IMC controller should satisfy following conditions.
(i) If the process G_p has unstable poles or poles near zero at $z_1, z_2, \cdots, z_m$, then q should have zeros at $z_1, z_2, \cdots, z_m$.
(ii) If the process G_D has unstable poles or poles near zero at $z_{d1}, z_{d2}, \cdots, z_{dm}$, then $1 - G_p q$ should have zeros at $z_{d1}, z_{d2}, \cdots, z_{dm}$.
Since the IMC controller q is designed as $q = p_m^{-1} f$, the first condition is satisfied automatically because p_m^{-1} is the inverse of the model portion with the unstable poles or poles near zero. The second condition can be fulfilled by designing the IMC filter f as:

$$f=\frac{\sum_{i=1}^{m}\alpha_i s^i+1}{(\lambda s+1)^r} \tag{5}$$

where λ is an adjustable parameter which controls the trade–off between the performance and robustness, r is selected to be large enough to make the IMC controller (semi–)proper, and α_i is determined by Eq. (6) to cancel the unstable poles or poles near zero in G_D.

$$1-G_p q\Big|_{s=z_{d1},\cdots z_{dm}}=\left|1-\frac{p_A(\sum_{i=1}^{m}\alpha_i s^i+1)}{(\lambda s+1)^r}\right|_{s=z_{d1},\cdots,z_{dm}}=0 \tag{6}$$

Then, the IMC controller is described as:

$$q=p_m^{-1}\frac{(\sum_{i=1}^{m}\alpha_i s^i+1)}{(\lambda s+1)^r} \tag{7}$$

Thus, the resulting set–point and disturbance responses are obtained as:

$$\frac{y}{r}=G_p q f_R=P_A\frac{\left(\sum_{i=1}^{m}\alpha_i s^i+1\right)}{(\lambda s+1)^r}f_R \tag{8}$$

$$\frac{y}{d}=\left(1-G_p q\right)G_D=\left(1-P_A\frac{\left(\sum_{i=1}^{m}\alpha_i s^i+1\right)}{(\lambda s+1)^r}\right)G_D \tag{9}$$

The numerator expression $\left(\sum_{i=1}^{m}\alpha_i s^i+1\right)$ in Eq. (8) causes an excessive overshoot in the servo response, which can be eliminated by introducing the set–point filter to compensate for the overshoot in the servo response.

From the above design procedure, a stable, closed–loop response can be achieved by using the IMC controller. The ideal feedback controller that is equivalent to the IMC controller can be expressed in terms of the internal model $\tilde{G}_p$ and the IMC controller q:

$$G_c=\frac{q}{1-\tilde{G}_p q} \tag{10}$$

Substituting Eqs. (4) and (7) into (10) gives the ideal feedback controller:

$$G_c=\frac{p_m^{-1}\dfrac{(\sum_{i=1}^{m}\alpha_i s^i+1)}{(\lambda s+1)^r}}{1-\dfrac{p_A\left(\sum_{i=1}^{m}\alpha_i s^i+1\right)}{(\lambda s+1)^r}} \tag{11}$$

Controller Design for SOPDT Process

On the basis of the above design principle, the second-order plus dead time (SOPDT) process has been considered as a representative model:

$$G_p=G_D=\frac{Ke^{-\theta s}}{(\tau_1 s+1)(\tau_2 s+1)} \tag{12}$$

where K is the gain, τ_1 and τ_2 are the time constants, and θ the time delay. The IMC filter structure is chosen as

$$f=\frac{\left(\alpha_2 s^2+\alpha_1 s+1\right)}{(\lambda s+1)^4} \tag{13}$$

The IMC filter form in Eq. (13) was also utilized by several researchers[2,4,6,12] to design the IMC–PID controller. The resulting IMC controller becomes

$$q=\frac{(\tau_1 s+1)(\tau_2 s+1)(\alpha_2 s^2+\alpha_1 s+1)}{K(\lambda s+1)^4} \tag{14}$$

Therefore, the ideal feedback controller is obtained as

$$G_c=\frac{(\tau_1 s+1)(\tau_2 s+1)(\alpha_2 s^2+\alpha_1 s+1)}{K\left[(\lambda s+1)^4-e^{-\theta s}(\alpha_2 s^2+\alpha_1 s+1)\right]} \tag{15}$$

Since the ideal feedback controller in Eq. (15) does not have the PID controller form, the remaining issue is to design the PID controller with a lead-lag compensator (i.e., PIDC controller) that approximates the ideal feedback controller most closely. The ideal feedback controller, G_c, equivalent to the IMC controller, can be obtained after the approximation of the dead time $e^{-\theta s}$ by 1/1 Pade expansion as

$$e^{-\theta s}=\frac{(1-\theta s/2)}{(1+\theta s/2)} \tag{16}$$

and results in

$$G_c=\frac{\left[(\tau_1 s+1)(\tau_2 s+1)\right](\alpha_2 s^2+\alpha_1 s+1)(1+\theta s/2)}{K(\theta+4\lambda-\alpha_1)s\left[1+\frac{(\alpha_1\theta/2-\alpha_2+2\lambda\theta+6\lambda^2)}{(\theta+4\lambda-\alpha_1)}s+\frac{(\alpha_2\theta^2/2+3\lambda^2\theta+4\lambda^3)}{(\theta+4\lambda-\alpha_1)}s^2+\frac{(2\lambda^3\theta+\lambda^4)}{(\theta+4\lambda-\alpha_1)}s^3+\frac{\lambda^4\theta^2/2}{(\theta+4\lambda-\alpha_1)}s^4\right]} \tag{17}$$

As seen in Eq. (17), the resulting controller has the form of a PID controller cascaded with a high-order filter. The analytical PID formula can be obtained by rearranging Eq. (17) to give

$$K_C=\frac{\alpha_1}{K(4\lambda+\theta-\alpha_1)};\quad \tau_I=\alpha_1;\quad \tau_D=\frac{\alpha_2}{\alpha_1} \tag{18}$$

Furthermore, it is obvious from Eq. (6) that the remaining part of the denominator in Eq. (17) contains the factor of the process poles $(\tau_1 s+1)(\tau_2 s+1)$. Therefore, the parameter b in Eq. (1) can be obtained by taking the first derivative of Eq. (19) below:

$$(cs^2+bs+1)=\frac{\left[1+\frac{(\alpha_1\theta/2-\alpha_2+2\lambda\theta+6\lambda^2)}{(\theta+4\lambda-\alpha_1)}s+\frac{(\alpha_2\theta^2/2+3\lambda^2\theta+4\lambda^3)}{(\theta+4\lambda-\alpha_1)}s^2+\frac{(2\lambda^3\theta+\lambda^4)}{(\theta+4\lambda-\alpha_1)}s^3+\frac{\lambda^4\theta^2/2}{(\theta+4\lambda-\alpha_1)}s^4\right]}{(\tau_1 s+1)(\tau_2 s+1)} \tag{19}$$

and substituting $s=0$ as

$$b=\frac{(\alpha_1\theta/2-\alpha_2+2\lambda\theta+6\lambda^2)}{(\theta+4\lambda-\alpha_1)}-(\tau_1+\tau_2) \tag{20}$$

The parameter a in Eq. (1) can be easily obtained from Eq. (17) as

$$a=0.5\theta \tag{21}$$

Since the high-order cs^2 term has little impact on the overall control performance in the control relevant frequency range, the remaining part of the fraction in Eq. (17) can be successfully approximated to a simple, first-order lead/lag compensator as $(1+as)/(1+bs)$.

The values of α_1 and α_2 are selected to cancel out the poles at $-1/\tau_1$ and $-1/\tau_2$. This requires $[1-Gq]\big|_{s=1/\tau_1,\,1/\tau_2}=0$ and thus $\left[1-(\alpha_2 s^2+\alpha_1 s+1)e^{-\theta s}/(\lambda s+1)^4\right]\big|_{s=-1/\tau_1,\,-1/\tau_2}=0$. The values of α_1 and α_2 are obtained after simplification and given below.

$$\alpha_1 = \frac{\tau_1^2\left[\left(1-\frac{\lambda}{\tau_1}\right)^4 e^{-\theta/\tau_1}-1\right]-(\tau_2)^2\left[\left(1-\frac{\lambda}{\tau_2}\right)^4 e^{-\theta/\tau_2}-1\right]}{(\tau_2-\tau_1)} \quad (22)$$

$$\alpha_2 = \tau_2^2\left[\left(1-\frac{\lambda}{\tau_2}\right)^4 e^{-\theta/\tau_2}-1\right]+\tau_2\alpha_1 \quad (23)$$

A set–point filter is suggested to enhance the servo response, which is comprised of the lag term $(\alpha_2 s^2+\alpha_1 s+1)$, by eradicating the excessive overshoot and f_R is given as:

$$f_R = \frac{\gamma\alpha_1 s+1}{(\alpha_2 s^2+\alpha_1 s+1)} \quad (24)$$

where $0\le\gamma\le1$. The extreme case with $\gamma=0$ has no lead term in the set–point filter which would cause a slow servo response. Note that γ can be adjusted online to obtain the desired speed of the set–point response.

In the simulation study, a widely accepted, weighting-type set–point filter is used for all methods, except for the proposed method, to eliminate the overshoot in the set–point response. This filter is given as:

$f_R = (\gamma\tau_I s+1)/(\tau_I\tau_D s^2+\tau_I s+1)$

where $0\le\gamma\le1$.

Regarding the classification of the process model to deal with the proposed design method, the unstable processes which do not have the form of Eq. (12) can be easily transformed to the form of Eq. (12) by adjusting their sign for controller design.

Remarks 1: For the second-order, stable/unstable process without any zero, the designed value of b is too large for the closed–loop system to obtain robust performances when the parametric uncertainties are large. Based on extensive simulation study that has been conducted on different stable/unstable processes, a robust control performance was achieved by using a value of $0.1b$ instead of b [11-12].

Simulation Study

Example 1. Stable SOPDT Process

Consider the following SOPDT process (Chen and Seborg[5], Shamsuzzoha and Lee[4]):

$$G_p = G_D = \frac{2e^{-1s}}{(10s+1)(5s+1)} \quad (25)$$

The proposed PIDC controller is compared with two other PID controllers: DS–d[5] and the method of Shamsuzzoha and Lee[4] (hereafter, SL method). The controller parameters, including the performance and robustness matrix, are listed in Table 1. In order to ensure a fair comparison, all of the compared controllers are tuned to have $Ms=1.87$ by adjusting their respective λ.

A unit step change is introduced in both the set–point and load disturbance. Figure 2 compares the set–point and load disturbance responses obtained using three compared controllers. The 2DOF control scheme using the set–point filter was used in each method to enhance the set–point response. The proposed controller shows both smaller overshoot and faster disturbance rejection than those by the DS–d and SL methods.

Although the overshoot is almost similar for the DS–d and SL methods, the DS-d method shows the slowest response and requires long settling time. The proposed controller shows significant advantages in overshoot and fast settling time, particularly in disturbance rejection. The closed–loop response for both the set–point tracking and disturbance rejection confirms the superior response of the proposed controller for the same robustness.

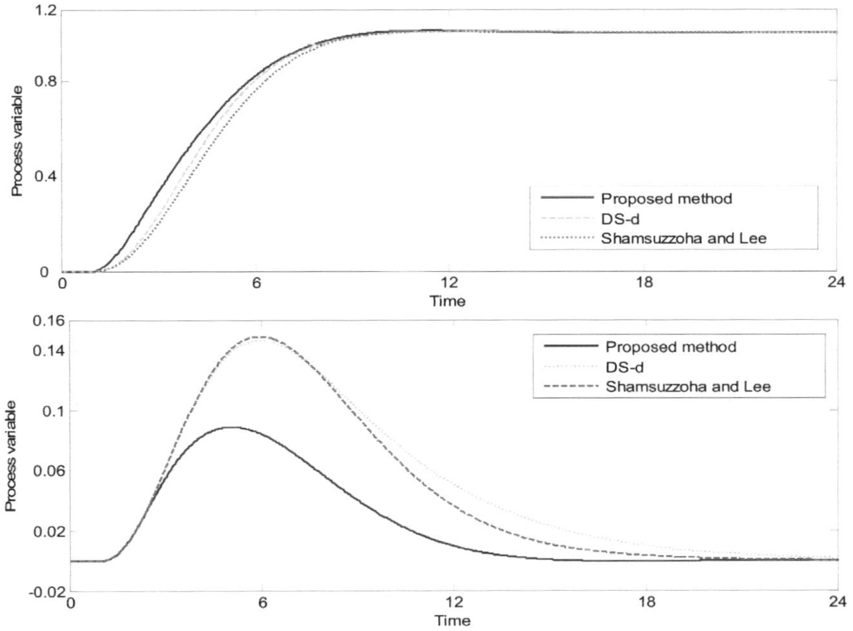

Figure 2. Response of the nominal system for Example 1.

Table 1. Controller parameters and resulting performance indices for Example 1

Method	K_c	τ_I	τ_D	set–point		disturbance	
				nominal case	10% mismatch	nominal case	10% mismatch
				ITAE	*ITAE*	*ITAE*	*ITAE*
Proposed $\lambda=1.182$	9.80	5.45	1.68	10.89	11.28	3.50	3.49
Chen and Seborg $\lambda=2.40$	6.38	7.60	2.09	12.68	13.02	9.77	9.70
Shamsuzzoha and Lee $\lambda=1.60$	6.41	6.85	1.97	13.17	13.97	7.90	7.88

– Proposed, $\gamma=0.3$, $f_R=(1.6351s+1)/(9.2099s^2+5.4502s+1)$, a=0.5, b=0.0341

– Chen and Seborg, $\gamma=0.5$, $f_R=(3.8022s+1)/(15.9520s^2+7.6045s+1)$

– Shamsuzzoha and Lee, $\gamma=0.4$, $f_R=(2.7437s+1)/(13.5800s^2+6.8593s+1)$

– Ms= 1.8695

The robust performance is evaluated by simultaneously inserting a perturbation uncertainty of 10% in all three parameters in the worst direction and finding the actual process as $G_p = G_D = 2.2e^{-1.1s}/(9s+1)(4.5s+1)$. The simulation results for the model mismatch for the three methods are given in Table 1. The performance and robustness indices clearly demonstrate the superior robust performance of the proposed controller.

Example 2: First-order Delay Integrating Process

Consider the first-order integrating process used by Zhang et al.[13] and Sree and Chidambaram[14].

$$G_p = G_D == \frac{1e^{-4s}}{s(4s+1)} \tag{26}$$

In the simulation study, we compare the proposed PIDC controller with the PIDC controller by Zhang et al.[13] and the PID controller by Sree and Chidambaram[14]. For both the proposed method and that of Zhang et al.[13], λ is adjusted to obtain $Ms = 3.28$ in order to ensure the same robustness with that of Sree and Chidambaram's[14] method.
The proposed controller was designed by considering the above process as $G_p = 100e^{-4s}/(100s+1)(4s+1)$. Figure 3 shows the closed–loop output response for a unit step change in both the set–point and load disturbance. The controller setting parameters and performance matrix for both the set–point and load disturbance are listed in Table 2. To eliminate the overshoot in the set–point response, a set–point filter is used in each method, as provided in Table 2. The proposed controller has a small peak and fast settling time in the disturbance rejection, whereas in the set–point it has no overshoot and a faster settling time than that of the other methods. This comparison of the output response and the values of the performance matrices listed in Table 2 confirms the superior performance of the proposed controller.

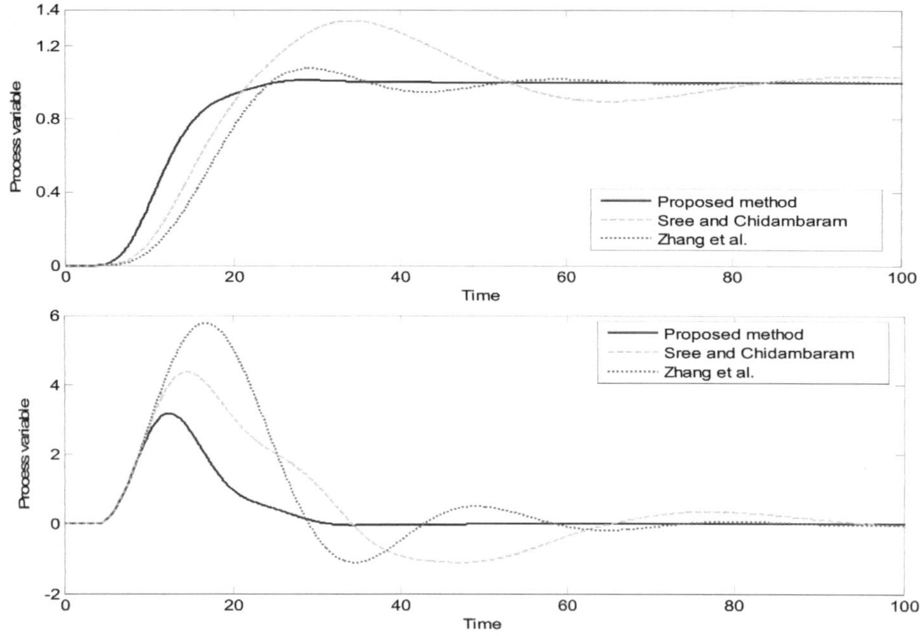

Figure 3. Response of the nominal system for Example 2.

To confirm the robust performance of the proposed controller, 10% parameter perturbations were assumed in K, τ and θ simultaneously towards the worst-case model mismatch and in finding the actual process as $G_p = G_D = 1.1e^{-4.4s}/s(3.6s+1)$. The simulation results for the model mismatch for various methods are also given in Table 2. The robust performance indices demonstrate the superior robust performance of the proposed controller.

Table 2. Controller parameters and resulting performance indices for Example 2

Method	K_c	τ_I	τ_D	set–point		disturbance	
				nominal case	10% mismatch	nominal case	10% mismatch
				ITAE	*ITAE*	*ITAE*	*ITAE*
Proposed $\lambda = 2.117$	0.3593	12.130	2.704	87.31	128.9	499.4	953.1
Sree and Chidambaram	0.1990	10.0	7.020	560.0	478.6	3092.0	2835.0
Zhang et al. $\lambda = 2.782$	0.2251	16.346	3.021	203.8	243.8	2109.0	3032.0

– Proposed, $\gamma = 0$, $f_R = 1/(32.8106s^2+12.1304s+1)$, a=2.0, b=0.049

– Sree and Chidambaram, $\gamma = 0$, $f_R = 1/(70.2s^2+10s+1)$

– Zhang et al., $\gamma = 0$, $f_R = 1/(49.3845s^2+16.346s+1)$, b= 0.2966

– Ms=3.2865

Example 3. SODUP (Two Unstable Poles)

The following widely published SODUP example has been considered for the comparison (Rao and Chidambaram[14], Liu et al.[9]):

$$G_p = G_D = \frac{2e^{-0.3s}}{(3s-1)(1s-1)} \tag{27}$$

The recently published papers of Rao and Chidambaram[14] and Liu et al.[9] demonstrated the superiority of their methods for the SODUP over several widely accepted, previous approaches. Therefore, the proposed method is compared with those of Rao and Chidambaram[14] and Liu et al.[9]. For a fair comparison, λ for the proposed method is adjusted to give the same $Ms = 3.09$ as for Rao and Chidambaram[14] and the tuning parameters are listed in Table 3. For the Liu et al.[9] method, a high-order controller is used for the comparison because they demonstrated a significant performance improvement by using the high-order controller. Figure 4 compares the proposed method with the Rao and Chidambaram[14] and Liu et al.[9] methods, by introducing a unit step change in the set–point and the load disturbance, respectively. For the servo response, the set–point filter is used for both the proposed and Rao and Chidambaram[14] methods, whereas the three-control-element structure is used for the Liu et al.[9] method. It is important to note that the well–known, modified IMC structure has the theoretical advantage of eliminating the time delay from the characteristic equation. Unfortunately, this advantage is lost if the process model is inaccurate.

Besides, real process plants usually incorporate unmodeled dynamics that inevitably tend to deteriorate the control system performance severely.
Fig. 4 and performance indices listed in Table 3 clearly show significantly improved load disturbance response and fast settling time of the proposed method. All three methods exhibit similar servo responses.

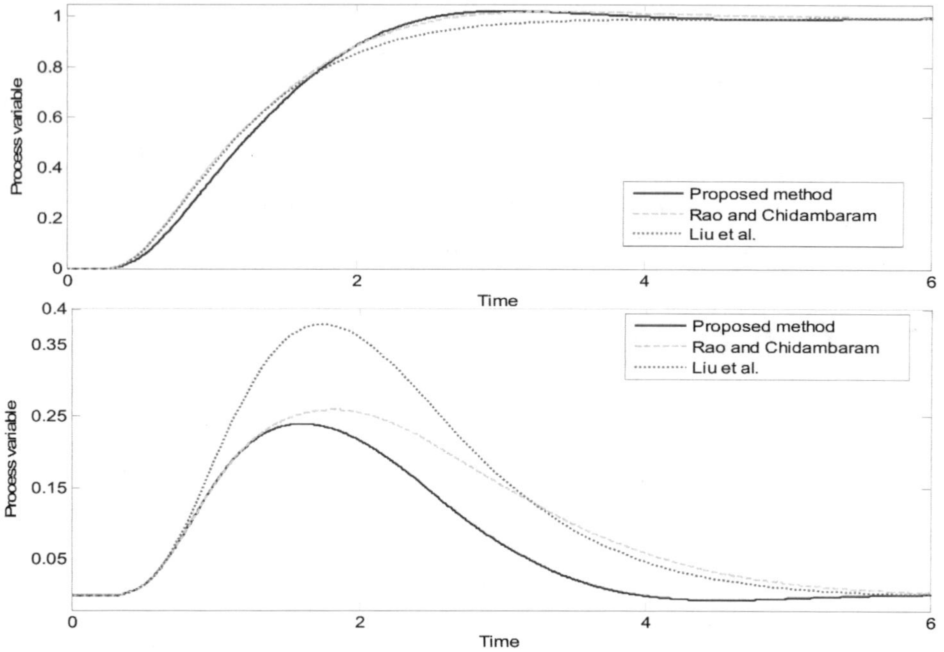

Figure 4. Response of the nominal system for Example 3.

Table 3. Controller parameters and resulting performance indices for Example 3

Method	K_c	τ_I	τ_D	set–point nominal case ITAE	set–point 5% mismatch ITAE	disturbance nominal case ITAE	disturbance 5% mismatch ITAE
Proposed $\lambda = 0.3555$	3.4706	1.5052	1.3633	1.103	1.254	0.86	0.901
Rao and Chidambaram $\lambda = 0.64$	2.9723	1.9003	1.6141	1.032	1.378	1.51	1.583
Liu et al. $\lambda = 0.51$	-	–	–	1.267	1.417	1.67	1.886

– Proposed, $\gamma = 0.3$, $f_R = (0.4516s+1)/(2.0519s^2+1.5052s+1)$, a=0.15, b= 0.0059

– Rao and Chidambaram, a set–point weighting parameter 0.536 is used, a=0.15, b= 0.0041

– Liu et al., two– degree–of–freedom control structure with three controller:

a disturbance estimator $G_c = \dfrac{32.82s^3 + 439.41s^2 + 232.64s + 129.79}{0.56s^3 + 0.8s^2 + 100s}$, a set–point controller $k_d = 3.0$, $C(s) = \dfrac{1.5s^2 + s + 0.5}{(0.51s+1)^2}$

– Ms= 3.0915

The robustness of the controller was investigated by inserting a perturbation uncertainty of 5% in all three parameters simultaneously towards the worst-case model mismatch, i.e., $G_p = G_D = 2.1e^{-0.315s}/(2.85s-1)(0.95s-1)$. The simulation results presented in Table 3 for both the set–point and the disturbance rejection clearly show the excellent set–point and load response of the proposed PIDC controller. The modified IMC controller corresponding to the Liu et al.[9] method has the worst performance for both set–point and disturbance rejection response for the model mismatch.

Remarks 2: Closed-loop time constant (λ) guidelines

The closed–loop time constant λ is the only user–defined tuning parameter in the proposed tuning rule. It is directly related to the performance and robustness of the proposed controller and it is important to set some λ guidelines in order to provide both a fast and robust performance. The small peak and fast response speed of disturbance rejection are favored by selecting a small value of λ; however, robustness and stability of the system require a sufficiently large value of λ. The extensive simulation studies presented here suggest that robust control performance can be achieved by considering the starting value of λ to be equal to the process time delay. If the selected λ does not satisfy the performance and robust stability criteria, then λ is monotonously increased or decreased on-line until the desirable trade–off between the close–loop nominal performance and its robust stability is achieved.

Conclusions

In this article, we have discussed a IMC based simple design method for a PIDC controller, i.e., a PID cascaded with a lead-lag compensator. The proposed method covered a broad class of second-order stable and unstable processes with time delay. Several important representative processes were considered in the simulation study in order to demonstrate the superiority of the proposed method. The design method was based on the disturbance rejection and a set–point filter was suggested to eliminate the overshoot in set–point response. The results showed that both nominal and robust performances of the PID controller can be significantly enhanced by adding a simple lead-lag compensator with a proper design method. The proposed controller consistently achieved superior performance for several process classes. In the robustness study conducted by simultaneously inserting a perturbation uncertainty in all parameters in order to obtain the worst-case model mismatch, the proposed method was found to be superior to the other methods.

Acknowledgement

The authors wish to thank the 2006 Energy Resource and Technology Project.

References

1 Morari M, Zafiriou E. Robust Process Control, NJ: Prentice–Hall Englewood Cliffs, 1989.
2 Lee Y, Park S, Lee M, Brosilow C. PID controller tuning for desired closed–loop responses for SI/SO systems. *AIChE J.* 1998;44:106–115.

3 Skogestad S. Simple analytic rules for model reduction and PID controller tuning. *J Process Control*. 2003;13:291–309.
4 Shamsuzzoha M, Lee M. IMC–PID controller design for improved disturbance rejection of time–delayed processes. *Ind Eng Chem Res.* 2007;46:2077–2091.
5 Chen D, Seborg DE. PI/PID controller design based on direct synthesis and disturbance rejection. *Ind Eng Chem Res.* 2002;41:4807–4822.
6 Lee Y, Lee J, Park S. PID controller tuning for integrating and unstable processes with time delay. *Chem Eng Sci.* 2000;55:3481–3493.
7 Yang XP, Wang QG, Hang CC, Lin C. IMC–based control system design for unstable processes. *Ind Eng Chem Res.* 2002;41:4288–4294.
8 Wang YG, Cai WJ. Advanced proportional–integral–derivative tuning for integrating and unstable processes with gain and phase margin specifications. *Ind Eng Chem Res.* 2002;41:2910–2914.
9 Liu T, Zhang W, Gu D. Analytical design of two–degree–of–freedom control scheme for open–loop unstable process with time delay. *J Process Control.* 2005;15:559–572.
10 Zhang WD, Gu D, Wang W, Xu X. Quantitative performance design of a modified smith predictor for unstable processes with time delay. *Ind Eng Chem Res.* 2004;43:56–62.
11 Rao AS, Chidambaram M. Enhanced two–degrees–of–freedom control strategy for second–order unstable processes with time delay. *Ind Eng Chem Res.* 2006;45:3604–3614.
12 Shamsuzzoha M, Jeon J, Lee M. Improved analytical PID controller design for the second order unstable process with time delay. 17th European Symposium on Computer Aided Process Engineering (ESCAPE-17), Bucharest, Romania, 27-30 May 2007 (Editors Valentin Plesu and Paul Serban Agachi), 901-906.
13 Zhang W, Xu X, Sun Y. Quantitative performance design for integrating processes with time delay. *Automatica.* 1999;35:719–723.
14 Sree RP, Chidambaram M. A simple and robust method of tuning PID controllers for integrating/dead time processes. *Journal of chemical Engineering of Japan.* 2005;38:113–119.

Studies on Local Mass Transfer with Orifice-Disc Turbulence Promoter

Ganapathi Rao, Y., Vaka Murali Mohan., Rajendra Prasad, P and Sarveswarao, S

Electro Chemical Engineering Research laboratory,
Department of Chemical Engineering,
Andhra University, Visakhapatnam-03, A P, INDIA.
E-mail : prprasad_au@yahoo.co.in

ABSTRACT

Augmentation of mass transfer rates becomes imperative for any process intensification. Limiting current data for forced convection flow of electrolyte through the conduit with orifice-disc assembly as turbulence promoter were obtained. The effect of geometric parameters like orifice diameter, disc diameter, and height of the disc from orifice was studied. Augmentation in mass transfer coefficients were found to vary from 3 to 15 fold depending upon flow rate and geometric parameters. Mass transfer coefficient increased with increase in disc diameter, decreased with increase in orifice diameter and orifice-disc distance. A correlation for mass transfer in terms of geometric parameters was developed and presented.

Keywords: disc, turbulence promoter, circular conduit, mass transfer,

INTRODUCTION

Augmentation of mass transfer rates become imperative for any transfer processes. become necessary either to reduce the equipment size or to increase throughputs. Essentially electrochemical processes like electro winning, electrorefming, electro milling, electroplating, etc. are diffusion controlled. These processes want high transfer rates. In a fluid various flow patron can be developed due to turbulence generated by the a) insert promoters of various geometry. b) vibration or rotations of the fluid or the transfer c) location of transfer element transverse to the now, d) impinging jets and e) fluidizing solid etc. Large number of studies has been reported in the literature employing the above techniques.

Sherwood, J D and Stone, H A [1] determine the flow of in viscid fluid around a disc in a pipe and the added mass of the accelerating disc and flow profiles are also presented. Ravi et al [2] did experiment and obtain mass transfer data on the walls of the fluidized beds using an electrochemical technique with a potassium ferri-ferro cyanide couple. The disc promoter assembly was used as internal in fluidized beds. Burns, J R and Jachuck, R J J [3] determined, Liquid–solid mass transfer performance on a 30 cm diameter spinning disc reactor by use of the limiting current technique. Podestá, J J et al [4] obtained a dimensionless correlation for ionic mass transfer at horizontal disc electrodes under longitudinal vibration. Escudié, R et al [5] carried out experiments in water-fluidized binary (*and ternary*) mixtures of teflon spheres, discs and rods. Venkateswarlu and Jagannadha Raju [6] studied the effect if string of discs placed co-axially in a circular conduit on momentum transfer. Venkateswarlu, P [7] obtained momentum and mass transfer data with co-axially placed discs on a central rod. The augmentation was about 12 fold over the data in conduits in the absence of disc promoter elements. Ziółkowska, I and Dolata, M [8] experimentally investigated the heat transfer between a gas stream and the wall of a tube with discs as turbulence promoters. Mishra and Prakash Chandra Singh [9] studied mass transfer for disks spinning under laminar and turbulent flow conditions. Prandtl analogy for turbulent flow mass transfer modified angle of attack. Sitaraman [10] studied augmentation of mass transfer with a string of spheres placed axially. He studied wall and surface mass transfer coefficients both with and without fluidizing solids. He obtained pressure drop data for the case of homogeneous flow of fluid electrolyte. Rajendra Prasad [11] studied ionic mass and momentum transfer with coaxially placed spiral coils as turbulent promoter in homogeneous and in fluidized beds and did experiment using ferro-ferri cyanide system by limiting current technique. Sujatha [12] studied mass and momentum transfer using helical tapes mounted on central rod as turbulence promoter. Redox reactions of ferri-ferro cyanide ions were selected as systems for her experimental study. Sarveswara Rao [13] employed co-axially placed cones on a central rod as promoter elements. The augmentation in case of cones was found to be more effective than in the streamlined bodies and discs. Twenty fold increase in mass transfer coefficient was observed over that of bare tube within the range of variables covered.

EXPERIMENTATION

The schematic diagram of the experimental unit was shown in Fig. 1. The equipment essentially consisted of a re-circulating tank, a centrifugal pump, two rotameters, an entrance calming section, a test section, and an exit calming section. The test section was provided with a number of point electrodes fixed flush with its surface longitudinally at equal spacing (0.01 m). The promoter was mounted in the test section coaxially by means of flanged joint. Coil turbulence promoter was made of copper wire with provision to fix it rigidly within test section. The details of the coil promoter were shown in Fig. 2. After inserting the promoter in the column, hundred liter of the electrolyte consisting of 0.01 M potassium ferricyanide,

0.01 M potassium ferrocyanide, and 0.5N sodium hydroxide was prepared in the storage tank. The electrolyte was pumped through the test section. By adjusting the flow rate of the electrolyte by operating the control and by pass valves. Limiting current data were measured at point copper electrodes for the reduction of potassium ferricyanide ion and for the oxidation of potassium ferrocyanide ion. The electrochemical reactions involved are given below:

$$Fe[(CN)_6]^{-3} + e^- \rightarrow Fe[(CN)_6]^{-4} \text{ for reduction,} \quad \text{(i)}$$
$$Fe[(CN)_6]^{-4} - e^- \rightarrow Fe[(CN)_6]^{-3} \text{ for oxidation} \quad \text{(ii)}$$

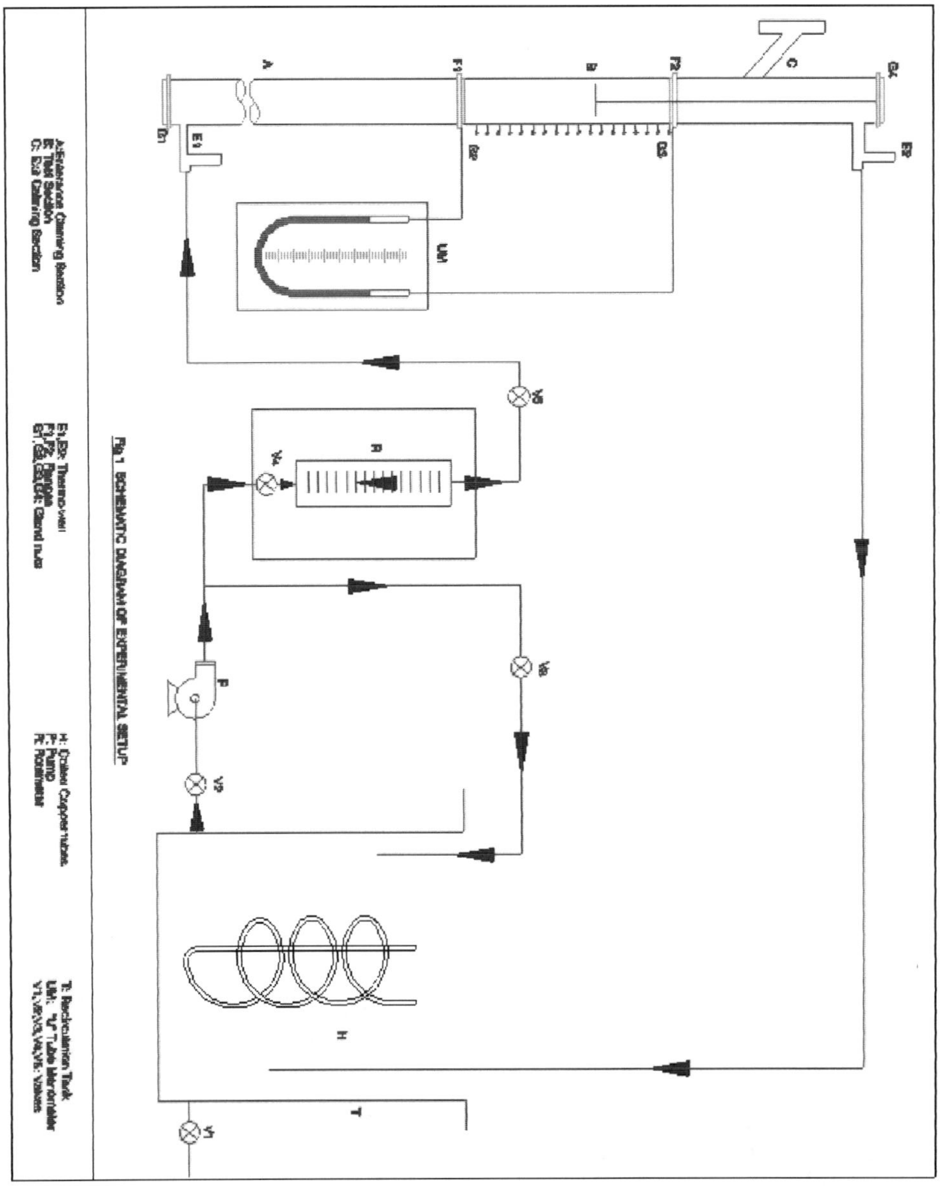

Fig 1 SCHEMATIC DIAGRAM OF EXPERIMENTAL SETUP

The limiting current was indicated by a small increase in current for a sharp increase in voltage. The experiments were repeated by varying the pitch of the spiral coil (P_c) and Length of the coil (L_c). Mass transfer coefficients are calculated by the following expression:

$$k_L = \frac{I_L}{n A F c_0} \qquad (iii)$$

The range of variables covered is given in Table 1.

Fig.2 Details of the Text Section

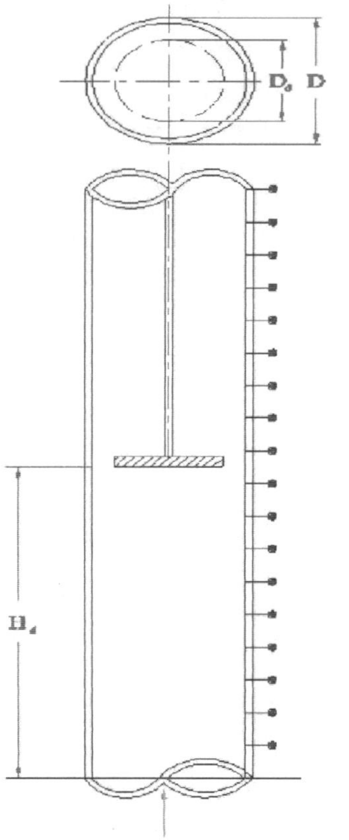

H_d : Height of the disc
D_d : Diameter of the disc
D : Diameter of the Conduit

RESULTS

Orifice disc placed across the flow causes obstruction to the flow masses the stream lines deviate from normal path and takes diversion in radial direction, takes axial become a creeping flow, and finally the flow again becomes normal flow. In taking this path, wakes and eddies are generated on the rear side of the disc, making the flow turbulent. When the disc is placed axially across the flow through pipes, the skin friction and form friction generated is even severe transform the flow more turbulent. In the present work, it is envisaged to harness the benefits generated out of the turbulence created by the disc when it is placed axially in pipe flow. Point electrodes are fixed flush with the surface of the wall of the test section was used for measuring the limiting current from these values mass transfer coefficients were computed. Observations of these mass transfer coefficient values reveal the effect of turbulence generated by the disc. This generated turbulence propagates in both the directions namely upstream and down stream directions. Therefore it is planed to identify the levels of turbulence in both directions. This is also directed to know the considerable augmentation in both directions by knowing the local mass transfer data. Once the extent of length is known in both directions, one can plan for the placement of other promoting element. The study is useful in transfer operations like reverse osmosis, electro winning, electro refining, electro organic synthesis and also be used as guide for mass transfer enhancements. It is envisage to obtain pressure drops to evaluate the study in terms of energy efficiencies.

Effect of disc diameter

Disc diameter has strong influence on mass transfer. Mass transfer coefficient vs velocity data was drawn and show as Fig.3. Mass transfer coefficient is increasing with velocity and with disc diameter. Data on Mass transfer coefficient vs disc diameter is generated from the above graph and shown in Fig.4. The disc has an exponent augmentation levels are varied with velocity, at velocity of 0.071 m/s augmentation varied from 5 to 15 folds, while that for 0.325 m/s varied from 3 to 10 folds. Disc is at higher velocities indicating the system has reached fully turbulent zone.

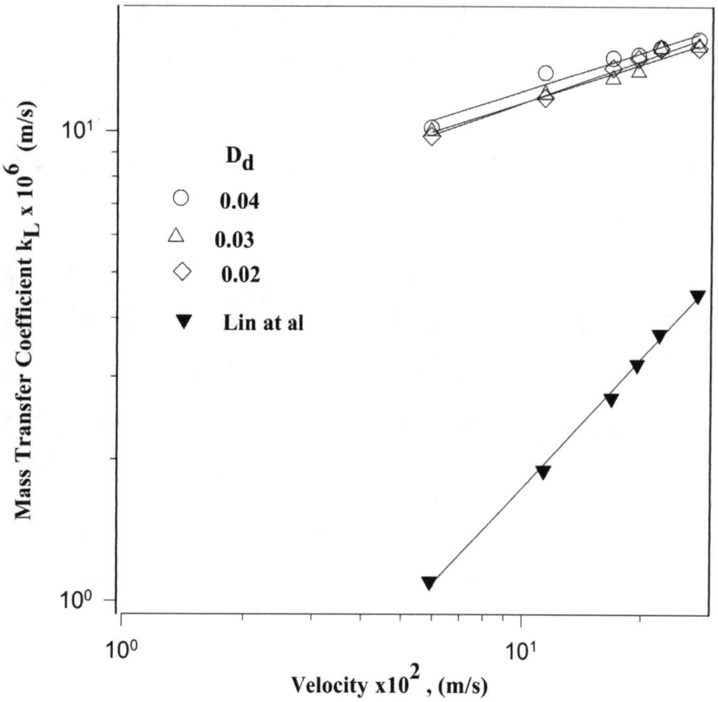

Fig. 3 Variation of mass transfer coefficient with velocity - effect of disc diameter on mass transfer coefficient

Effect of height of the disc

As height of the disc varied the extent of turbulence is also changes because of change in circulation patterns and extent in both regions. Mass transfer coefficients vs velocity data is plotted and shown as Fig.8. The figure reveals that mass transfer coefficient is increasing with increase in height of the disc. As height of the disc increases there is a decrease in down stream section there by the reduction in turbulent zone causes a decrease in average mass transfer coefficient. Augmentation are varied with velocity, at velocity of 0.071 m/s augmentation varied from 3 to 6 folds, while that for 0.325 m/s varied from 3 to 10 folds.

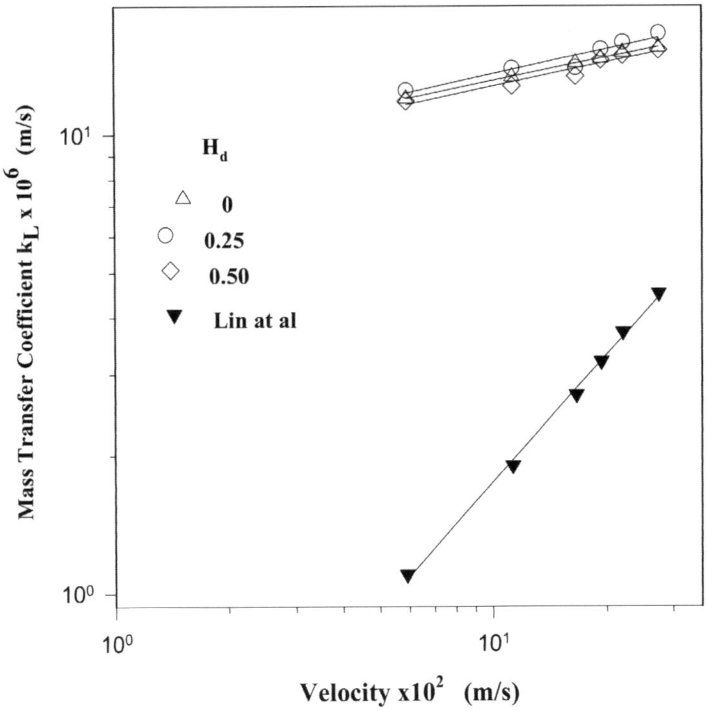

Fig.4 Variation of mass transfer coefficient with velocity - effect of height of the disc on mass transfer coefficient

CONCLUSIONS

1. Mass transfer coefficient is increasing with velocity,
2. Mass transfer coefficient increases with increase in disc diameter.
3. The height of the disc increases to maximum and then showed a decrease. 0.25 m height of the disc was suggested as it presents maximum augmentation.
4. Augmentations achieved, from 3 to 15 folds over smooth tube flow.

REFERENCES

1. Sherwood, J D and Stone, H A, "Added mass of a disc accelerating within a pipe" *Physics of Fluids -- Nov 1997 – Vol. 9, Iss 11, pp. 3141-3148*

2.. Ravi, T., Srinivasa Rao, B., Gopala Krishna, P and Venkateswarlu, P, "Ionic MT studies in fluidized beds with coaxially placed discs on a rod as internal" *Chemical Engineering and Processing Vol 35, Iss 3, June 1996, pp 187-193.*

3. Burns, J R and Jachuck, R J J, "Deter of liquid solid M Tcoeff for a spinning disc reactor using a limiting current technique" *Int. J of Heat and Mass Transfer Vol 48, Iss 12, June 2005, pp 2540-2547.*

4.Podestá, J J., Paús, G F and Arvía, A J, "Ionic mass transfer at horizontals disc electrodes under longitudinal vibration" *Electrochimica Acta Vol 19, Iss 9, Sep 1974, pp 583-589.*

5. Escudié, R., Epstein, N., Grace, J R and Bi, H T, "Effect of particle shape on liquid-fluidized beds of binary (*and ternary*) solids mixtures" *Chemical Engineering Science Vol*

61, Iss 5, March 2006, pp 1528-1539.
6. Ventateswarulu, P and Jagannadha Raju, G.J.V, *Indian Chemical Engineer Vol. 32, No.13, 101, 1990.*
7.Venkateswarlu, P, "Studies on ionic MT with coaxially placed discs on a rod as turbulence Promoter" *Ph.D., Thesis, Andhra University, Visakhapatnam, INDIA, 1987.*

7. R.G. Deissler, C.S. Eian, "Analytical and experimental investigation of adiabatic turbulent flow in smooth tubes" , *Nat. Adv. Comm. Aeronaut., Washington, DC, TN-2629, 1952.*
8. S.L. Lyons, T.J. Hanratty, J.B. McLaughlin, "Direct numerical simulations of passive heat transfer in turbulen channel flow", *Int. J. Heat and Mass Transfer 34, 1991, pp 1149- 1158.*
9. G.A. Hughmark, "Momentum, heat, and mass transfer analogy for turbulent flow in circular pipes", *Ind. Eng. Chem. Fundamentals 8 , 1969, pp 31-35.*
10. R.G. Deissler, "Analysis of turbulent heat transfer, mass transfer and friction in smooth tubes at high Prandtl and Schmidt". *Nat. Adv. Comm. Aeronaut., Washington, DC, TN-1210, 1955.*
11. E.M. Sparrow, J.A. Malina, "Variable property, constant-property, and entrance-region HT results for turbulent flow of water and oil in a circular tube", *Chem. Eng. Sci. 19, 1964, 953-962.*
12. D.F.Dipprey and R.H.Sabersky, "Heat and momentum transfer in smooth and rough tubes at various Prandtl numbers" *Int J of Heat and Mass Transfer Vol 6, 1963, pp 329-353.*
13. J.Nikuradse, "Laws of flow in rough pipes" *VDI Forsch, 361, 1933.*

NOMENCLATURE

d = diameter of the conduit, m
d_d = disc diameter, m
h_d = height of the disc, m
L = length of the conduit, m
g_c = gravity constant, m/s^2
V = velocity of the fluid, m/s
Dimensionless
Rem = Roughness Reynolds number
Re =Reynolds number, $dV\rho/\mu$
f = Friction factor
Greek symbols
ρ = density of the fluid, kg/m^3
μ = viscosity of the fluid, poise
δ = boundary layer thickness
ν = kinematics viscosity

Table.1 Range of variables covered in the present study

Variable	Minimum	maximum
Disc diameter, m	0.020	0.040
Height of the disc, m	0	0.50
Velocity, m/s	0.071	0.325
Reynolds Number, Re	1400	18,000

66b Single-Phase Rimming Flow Of Liquid In A Rotating Horizontal Cylinder

Saravanan Suppiah Singaram and Roshan JJ Jachuck. Chemical and Bio-molecular Engineering, Clarkson University, 8 Clarkson Avenue, PO Box 5707, Potsdam, NY 13699

Liquid flow with a free surface inside a rotating horizontal cylinder is classified as rimming flow and the flow is generated due to external forces associated with inertia, viscosity, gravity and surface tension. This paper will discuss the hydrodynamics of thin film flow inside a rotating horizontal cylinder. Process Intensification and Clean Technology (PICT) group has been developing thin film reactors technology based on the rotating tube concept. The main aim of this study is to develop fluid dynamic understanding that will help predict liquid film thickness, residence time and heat and mass transfer intensities. Liquid film thickness is measured using an interferometric technique. There is extensive experimental as well as theoretical data available on rimming flow inside a partially-filled rotating horizontal cylinder (confined geometry) and there is no published data on rimming flow studies with a continuous flow system.

Mass Transfer Enhancement of Turbulent Flow in Circular Conduits

Vaka Murali Mohan and Rajendra Prasad P*

Electro Chemical Engineering Research laboratory,
Department of Chemical Engineering,
Andhra University, Visakhapatnam-03, A P, INDIA.
E-mail : prprasad_au@yahoo.co.in

ABSTRACT

Augmentation of mass transfer rates due to the presence of coaxially placed discs inserts in the flow of electrolyte were presented in this paper. local limiting current data were obtained at various flow rates using promoters of varying disc diameter and height of the disc. Wide range of variables are covered, disc diameter varied as 0.02, 0.03, 0.04 m and height of the disc varied as 0, 0.25, 0.50 m. Disc inserts have significant effect on the enhancement of mass transfer, The augmentation varied from 1 to 6 folds over tube with out promoter.

Keywords: disc, turbulence promoter, circular conduit, mass transfer,

INTRODUCTION

Discs placed across the flow of circular conduits generate wakes and internal circulation within the column. The disturbance continuous to show its effect on either side of the disc and continue to a longer distance which in turn influences the wall mass transfer coefficient over the length of the column. The effectiveness of a disc varies with the length of the column under consideration. It is envisaged to find the influence of disc on mass transfer coefficient on either side of the disc over the column length. Such studies were not found in literature. Hence the study was conducted to know the effectiveness of disc in augmenting mass transfer coefficient. Sherwood, J D and Stone, H A [1] determine the flow of in viscid fluid around a disc in a pipe and the added mass of the accelerating disc and flow profiles are also presented. Ravi et al [2] did experiment and obtain mass transfer data on the walls of the fluidized beds using an electrochemical technique with a potassium ferri-ferro cyanide couple. The disc promoter assembly was used as internal in fluidized beds. Burns, J R and Jachuck, R J J [3] determined, Liquid–solid mass transfer performance on a 30 cm diameter spinning disc reactor by use of the limiting current technique. Escudié, R et al [4] carried out experiments in water-fluidized binary (*and ternary*) mixtures of teflon spheres, discs and rods. Venkateswarlu and Jagannadha Raju [5] studied the effect if string of discs placed co-axially in a circular conduit on momentum transfer. Venkateswarlu, P [6] obtained momentum and mass transfer data with co-axially placed discs on a central rod. The augmentation was about 12 fold over the data in conduits in the absence of disc promoter elements.

EXPERIMENTATION

The schematic diagram of the experimental unit was shown in Fig.1.The equipment essentially consisted of a re-circulating tank, a centrifugal pump, two Rota meters, an entrance calming section, a test section, and an exit calming section. The test section was provided with a number of point electrodes fixed flush with its surface longitudinally at equal spacing (0.03 m). The promoter was mounted in the test section coaxially by means of gland nuts. Disc turbulence promoter was made of copper with provision to fix it rigidly within test section.

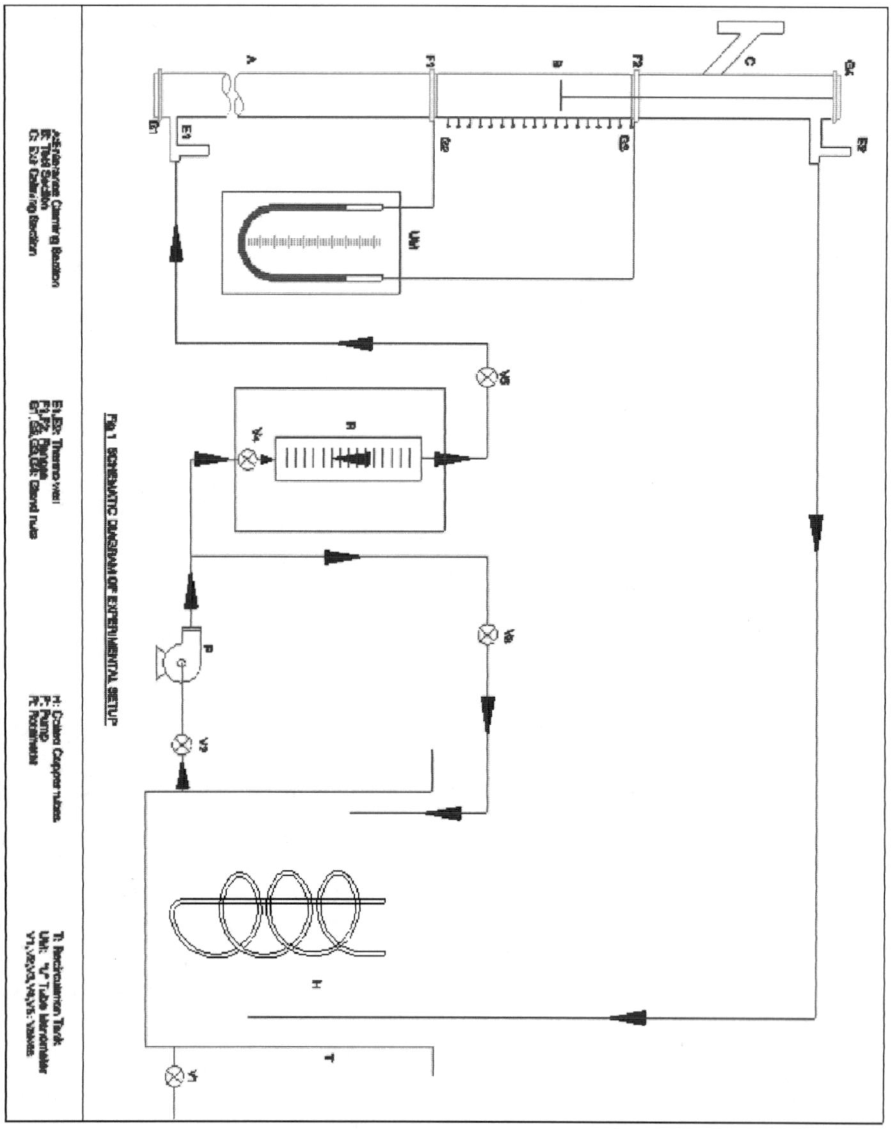

Fig 1. SCHEMATIC DIAGRAM OF EXPERIMENTAL SETUP

The details of the disc promoter were shown in Fig. 2. After inserting the promoter in the column, hundred liter of the electrolyte consisting of 0.01 M potassium ferricyanide, 0.01 M potassium ferrocyanide, and 0.5N sodium hydroxide was prepared in the storage tank. The electrolyte was pumped through the test section. By adjusting the flow rate of the electrolyte by operating the control and by pass valves. Limiting current data were measured at point copper electrodes for the reduction of potassium ferricyanide ion and for the oxidation of potassium ferrocyanide ion. The electrochemical reactions involved are given below:

$$Fe[(CN)_6]^{-3} + e^- \rightarrow Fe[(CN)_6]^{-4} \text{ for reduction,} \quad (i)$$

$$Fe[(CN)_6]^{-4} - e^- \rightarrow Fe[(CN)_6]^{-3} \text{ for oxidation} \quad (ii)$$

Fig.2 Details of the Text Section

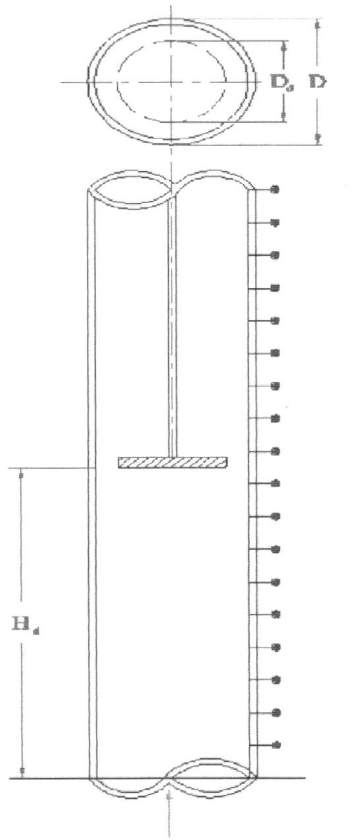

H_d: Height of the disc
D_d : Diameter of the disc
D : Diameter of the Conduit

The limiting current was indicated by a small increase in current for a sharp increase in voltage. The experiments were repeated by varying the disc diameter (d_d) and height of the disc (h_d). Mass transfer coefficients are calculated by the following expression:

$$k_L = \frac{I_L}{n\,A\,F\,c_0} \qquad (iii)$$

The range of variables covered is given in Table 1.

THEORY

Flow of electrolyte through tubular flow reactor with entry region coil insert promoters will establishes momentum and concentration profiles. The profile is divided into two regions, the inner region and outer region. The inner region again divided into laminar sub layer and viscous buffer zone. The outer region is turbulent core region and is fully turbulent. The turbulent core separated by an interface. A model is developed with the following assumptions:

1. The fluid is Newtonian, incompressible, and has constant transport properties and are independent of time.

2. The turbulent flow through smooth tube is fully developed with no entrance effects. The effect of body forces is small in comparison to that of viscous and inertial forces.

3. The axial velocity profile is independent of axial coordinate, similarly the concentration gradients.

4. The mass flux varies in the radial direction. The mass flux varies linearly with y, and is represented by the equation.

$$\frac{N}{N_w} = 1 - \frac{y}{R} \qquad 1$$

The radial variation of mass flux is taken into account by dividing into viscous-buffer zone and turbulent core. The flux at the interface is considered radially opposite directions, Flux at the interface towards the wall,

$$N_{i(v-b)} = k_{(v-b)}\,(c_i - 0) \qquad 2$$

Flux at the interface towards the axis,

$$N_{i(t)} = -k_{(t)}\,(c_i - c_b) \qquad 3$$

5. The shear stress is constant in the radial direction. Deissler and Eian [7] have found that the velocity and concentration profiles are relatively insensitive to shear stress distribution for fully developed turbulent fluid flow, hence can be assumed constant.

6. In the viscous buffer zone, the eddy diffusivity and eddy viscosity ratio varies along with the distance from the wall and is independent of Sc. Direct numerical simulations presented by Lyons et al. [8] suggest Sc to be unity at the wall. Hence, in the present case it is assumed that as $y^+ \rightarrow 0$, Sc=1.

7. The eddy diffusivity controls transfer process in viscous-buffer zone, while eddy viscosity controls the transport in turbulent core.

Development of model

In turbulent core, the intensity of eddies will be higher than the v-b zone. Therefore, during the transfer process, the resistance to the exchange of mass between the bulk fluid and the solid surface is often confined to the viscous and buffer region. The total resistance to the transfer process is the sum of the resistances in the viscous-buffer zone and turbulent layer, represented by,

$$\frac{1}{k}=\frac{1}{k_{v-b}}+\frac{1}{k_t} \qquad 4$$

Eq. (4) represents a two-layer resistance separated by an interface, similar to that proposed by Hughmark [9]. The total resistance in the v-b zone and the turbulent core is represented by Eq. (4) and are similar to the boundary layer resistance and the resistance corresponding to random eddies in the region beyond the boundary layer of Hughmark [9]. For a fully developed turbulent flow, the shear stress and the heat flux can be written as,

$$\tau=(\mu+\rho\varepsilon)\frac{du}{dy}=(\nu+\varepsilon)\frac{du}{dy} \qquad 5$$

$$N=(D+D_e)\frac{dc}{dy} \qquad 6$$

In the viscous-buffer region, Eq. (6) on integration with the boundary conditions:

At y=0 , c = 0 and at $y = y_i^+$, $c = c_i$

$$\frac{dc}{N_{i(v-b)}}=\int_0^{y_1^+}\frac{dy}{D+D_e}$$

$$\frac{dc}{N_{i(v-b)}}=\int_0^{y_1^+}\frac{\nu}{\sqrt{\tau_0/\rho}}\frac{dy^+}{\left[\frac{\nu}{Sc}+D_e(y^+)\right]}$$

$$\frac{\Delta c}{N_{i(v-b)}}=\frac{1}{\sqrt{\tau_0/\rho}}\int_0^{y_1^+}\frac{dy^+}{\frac{1}{Sc}+\frac{D_e(y^+)}{\upsilon}} \qquad 7$$

Substitute Equ. 2 in 7 yields,

$$\frac{1}{k_{v-b}}=\frac{1}{\sqrt{\tau_0/\rho}}\int_0^{y_1^+}\frac{dy^+}{\frac{1}{Sc}+\frac{D_e(y^+)}{\upsilon}} \qquad 8$$

As the transport is mainly by eddy means, the molecular diffusivity and the viscosity can be neglected in the turbulent core. Combining Eqs. 5 and 6 and integrating along the turbulent layer with the boundary conditions:

$c=c_b$ at $u = u_b$ and $c=c_i$ at $u = u_i$, yields,

$$\frac{dc}{du} = \frac{N_{i(t)}}{\tau_0}\left(\frac{\nu_e}{D_e}\right)_t \qquad 9$$

Substitute Equ. 3 in 9

$$\frac{dc}{du} = \frac{-k_t\, dc}{\tau_0}\left(\frac{\nu_e}{D_e}\right)_t$$

On rearrangement

$$\frac{1}{k_t} = \int_{y_1}^{0}\frac{-du}{\tau_0}\left(\frac{\nu_e}{D_e}\right)_t = \frac{-(u_i-u_b)}{\tau_0}.\Phi \qquad 10$$

where $\Phi = \left(\frac{\nu_e}{D_e}\right)_t$

$$\frac{1}{k_t} = \frac{(u_b-u_i)}{\tau_0}.\Phi = \frac{\Phi u_b}{\tau_0}\left(1-\frac{u_i}{u_b}\right)$$

$$\frac{1}{k_t} = \frac{\Phi u_b}{\tau_0}\left(1-\frac{u_i/u^*}{u_b/u^*}\right) = \frac{\Phi u_b}{\tau_0}\left(1-\frac{u_i^+}{u_b^+}\right) \qquad 11$$

When $u_m \cong u_b$

$$u_m^+ = u_b^+ = \sqrt{\frac{2}{f}} \qquad 12$$

Substitute Equ. 12 in 11

$$\frac{1}{k_t} = \frac{\Phi u_b}{\tau_0}\left(1-u_i^+\sqrt{f/2}\right) \qquad 13$$

Equ. 8 and 13 can be added using the resistance in series definition i.e. Equ. 4

$$\frac{1}{k} = \frac{1}{\sqrt{\tau_w/\rho}}\int_0^{y_1^+}\frac{dy^+}{\frac{1}{Sc}+\frac{D_e(y^+)}{\upsilon}} + \frac{\Phi u_b}{\tau_w}\left(1-u_i^+\sqrt{f/2}\right) \qquad 14$$

Mass transfer is aided by momentum flux or the molecular flux is negligible compared to turbulent flux. mass and momentum fluxes can taken equal and unity ($\Phi = 1$) and $\tau_w = 1$,

$$\frac{1}{k}=\int_0^{y_1^+}\frac{dy^+}{\frac{1}{Sc}+\frac{D_e(y^+)}{\upsilon}}+\sqrt{2/f}-u_i^+ \qquad 15$$

Substitute Equ. 3 in 15

$$\frac{c_b-c_i}{N_{it}}=\int_0^{y_1^+}\frac{dy^+}{\frac{1}{Sc}+\frac{D_e(y^+)}{\upsilon}}+\sqrt{2/f}-u_i^+$$

Where $c_b - c_i = \sqrt{f/2}/St$ and $N_{it}=1$

$$\sqrt{f/2}/St=\int_0^{y_1^+}\frac{dy^+}{\frac{1}{Sc}+\frac{D_e(y^+)}{\upsilon}}+\sqrt{2/f}-u_i^+$$

$$\sqrt{f/2}/St-\sqrt{2/f}=\int_0^{y_1^+}\frac{dy^+}{\frac{1}{Sc}+\frac{D_e(y^+)}{\upsilon}}-u_i^+$$

$$\frac{f/2St-1}{\sqrt{f/2}}=\int_0^{y_1^+}\frac{dy^+}{\frac{1}{Sc}+\frac{D_e(y^+)}{\upsilon}}-u_i^+$$

$$\frac{f/2St-1}{\sqrt{f/2}}+u_i^+=\int_0^{y_1^+}\frac{dy^+}{\frac{1}{Sc}+\frac{D_e(y^+)}{\upsilon}}=g(\mathrm{Re}_m^+,Sc)$$

$$\frac{f/2St-1}{\sqrt{f/2}}+u_i^+=g(\mathrm{Re}_m^+,Sc) \qquad 16$$

The function g is written as product of $\bar{g}$ and F(Sc).

$$g(\mathrm{Re}_m^+,Sc)=\bar{g}(\mathrm{Re}_m^+)F(Sc) \qquad 17$$

Re_m^+ is modified Reynolds number and is represented by

$$\mathrm{Re}_m^+=\frac{d_d}{d}\mathrm{Re}\sqrt{f/2}$$

The law of wall can also represent well for the present study also. It is assumed to be independent of Reynolds number. Sc is also assumed to be independent of Re_m^+ in the present study. The integral of Equ. 16 is only dependent on Sc only and is represented by F(Sc). The F(Sc) expression obtained in Prandtl and Von Karman anology is chosen for the present study. It is also supported by Deissler [10] and Sparrow [11] for Sc > 1. the empirical equation of F(Sc) is

$$F(Sc) = 9.3(Sc - 1)Sc^{-1/4} \quad 18$$

The function $\bar{g}$ is influenced by Re_m^+ and geometric parameters, .

$$\bar{g}(Re_m^+, \Phi_1, \Phi_2)$$

The resultant function can be represented as

$$\bar{g} = C\, Re_m^{+\,n_1}, \Phi_1^{\,n_2}, \Phi_2^{\,n_3}, \Phi_3^{\,n_3} \ldots\ldots\ldots\ldots\ldots\ldots\ldots\ldots \Phi_n^{\,n_{n-1}}$$

On regression analysis of present data yielded the following equation with deviations for disc insert is mentioned here under.

$$\bar{g} = 4593.4\, [Re_m^+]^{-0.00127} (\phi_1)^{-0.37357} (\phi_2)^{-0.01602} \quad 19$$

Average deviation = 3.7944, Standard deviation= 4.2555

RESULTS

Disc placed across the flow causes obstruction to the flow masses the stream lines deviate from normal path and takes diversion in radial direction, takes axial become a creeping flow, and finally the flow again becomes normal flow. In taking this path, wakes and eddies are generated on the rear side of the disc, making the flow turbulent. When the disc is placed axially across the flow through pipes, the skin friction and form friction generated is even severe transform the flow more turbulent. In the present work, it is envisaged to harness the benefits generated out of the turbulence created by the disc when it is placed axially in pipe flow. Point electrodes are fixed flush with the surface of the wall of the test section was used for measuring the limiting current from these values mass transfer coefficients were computed. Observations of these mass transfer coefficient values reveal the effect of turbulence generated by the disc. This generated turbulence propagates in both the directions namely upstream and down stream directions. Therefore it is planed to identify the levels of turbulence in both directions. This is also directed to know the considerable augmentation in both directions by knowing the local mass transfer data. Once the extent of length is known in both directions, one can plan for the placement of other promoting element. The study is useful in transfer operations like reverse osmosis, electro winning, electro refining, electro organic synthesis and also be used as guide for mass transfer enhancements. It is envisage to obtain pressure drops to evaluate the study in terms of energy efficiencies.

Local variation of mass transfer

Presence of disc in a circular conduit has strong influence on mass transfer. When a disc is placed in a column the column can be divided into two regions. The region approaching the disc is called upstream region, while that after the disc is called down stream region. The flow circulation and flow patterns are different for both region and hence behave differently. A graph drawn as mass transfer coefficients vs distance along the length of the column and shown as Fig.3. the disc placed at the starting of test section, the entire length of the test section behaves like a down stream region. Mass transfer coefficients are increasing with distance but the increase is maximum in the vicinity of disc. As we observe along the length of the column, the mass transfer coefficients is decreasing exponentially and reach steady state value. But is above the values of lin et al. further increase in length one could see a merger of local value those of lin et al, If the study is made at an extend length one could see the merger of local values with those of lin et al. In general mass transfer coefficients are increasing with velocity.

If the disc is placed at a distance of 0.25 m from the start of the test section, it could place in the middle of the test section. When local mass transfer coefficient values obtained from the study, constitute equal upstream and down stream regions. Augmentations observed would be maximum with this arrangement, as the turbulence is high around the disc. To that effect, figure is drawn and shown as Fig.4 it reveals the decaying sine wave present in both sides of the disc.

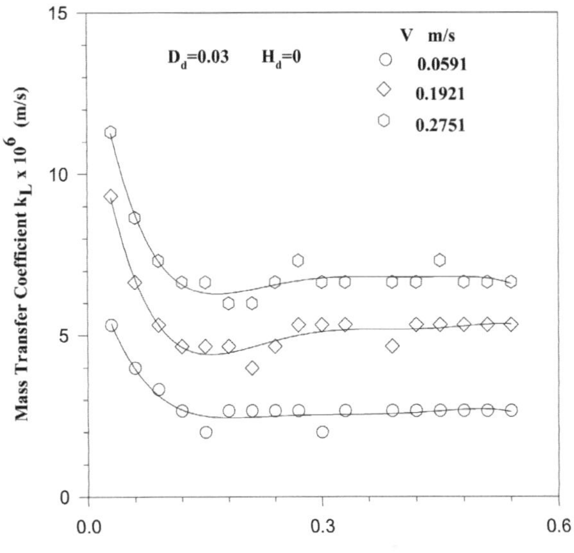

Fig.3 Variation of k_L with distance along the length of the conduit

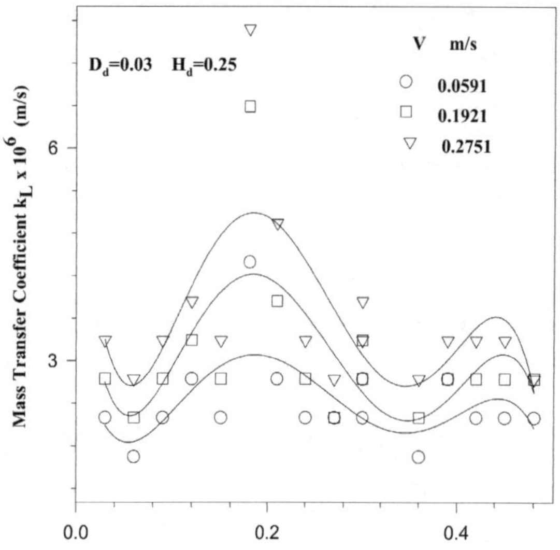

Fig.4 Variation of k_L with distance along the length of the conduit

Fig.5 is a graph between mass transfer coefficients and distance along the length of the conduit with velocity as parameter, in this case, the disc is placed at distance of 0.50 m from the start of the test section, the entire test section falls in upstream region. As the fluid approaches and collide with disc shock waves are generated, the wave has backward propagation resulting in decaying sine wave and results in increased turbulence yielded good augmentation and may continue further upstream. The length of upstream which has influence on mass transfer could be identified, if further study is made.

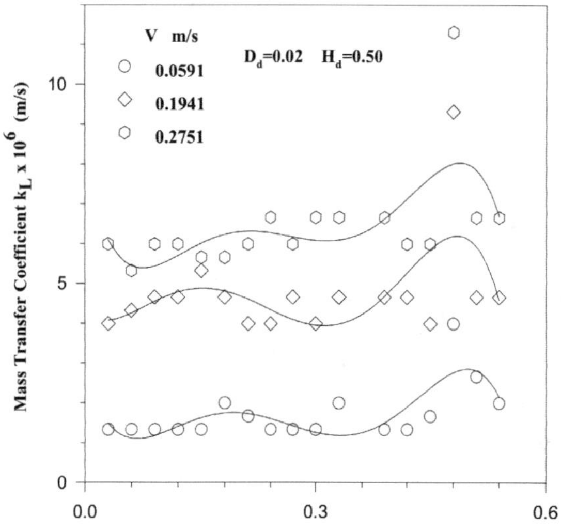

Fig.5 Variation of k_L with distance along the length of the conduit

Effect of disc diameter

Disc diameter has strong influence on mass transfer. Mass transfer coefficient vs velocity data was drawn and show as Fig.6. Mass transfer coefficient is increasing with velocity and with disc diameter. Data on Mass transfer coefficient vs disc diameter is generated from the above graph and shown in Fig.7. The disc has an exponent augmentation levels are varied with velocity, at velocity of 0.0591 m/s augmentation varied from 3 to 6 folds, while that for 0.2751 m/s varied from 1 to 2 folds. Disc is at higher velocities indicating the system has reached fully turbulent zone.

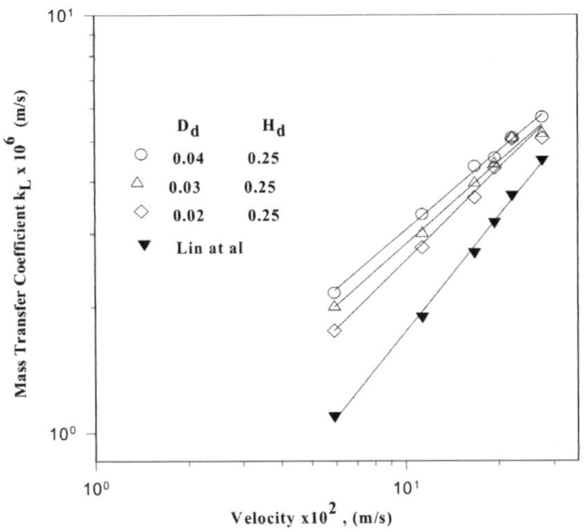

Fig. 6 Variation of mass transfer coefficient with velocity - effect of disc diameter on mass transfer coefficient

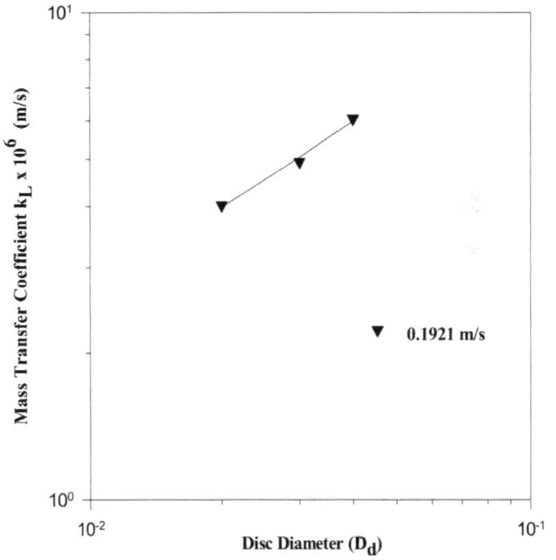

Fig. 7 Effect of disc diameter on mass transfer coefficient

Effect of height of the disc

The distance between the discs from the starting point of the test section is termed as the height of the disc. The height of the disc is varied from 0, 0.25, 0.50 m. the disc height of 0 m means, the entire section fall in the down stream region. If the height of the disc is 0.5 m the entire section falls in the up stream region. When height of the disc is 0.25 m the section is equally divided in to two regions.

As height of the disc varied the extent of turbulence is also changes because of change in circulation patterns and extent in both regions. Mass transfer coefficients vs velocity data is plotted and shown as Fig.8. The figure reveals that mass transfer coefficient is increasing with increase in height of the disc. As height of the disc increases there is a decrease in down stream section there by the reduction in turbulent zone causes a decrease in average mass transfer coefficient. Augmentation are varied with velocity, at velocity of 0.0591 m/s augmentation varied from 3 to 6 folds, while that for 0.2751 m/s varied from 1 to 2 folds. The variation of mass transfer coefficient with height of the disc is drawn and shown in Fig. 9, figure reveals that 0.25 m height is suggested to obtain good results.

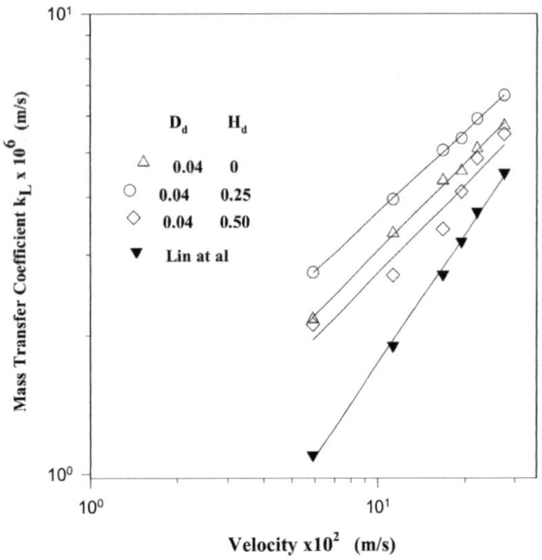

Fig.8 Variation of mass transfer coefficient with velocity - effect of height of the disc on mass transfer coefficient

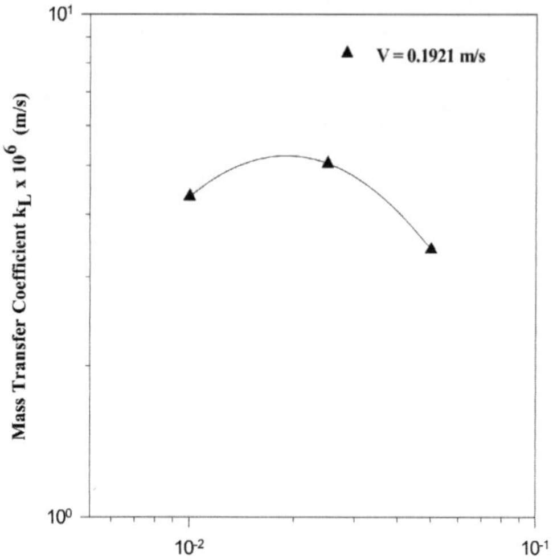

Fig.9 Effect of Height of the disc on mass transfer coefficient

Augmentations

A graph drawn augmentation factor vs velocity and shown in Fig.10. The graph reveals that the augmentation factor decreasing with velocity and increases with diameter of the disc. The augmentations achieved 1 to 6 folds over smooth pipe flow.

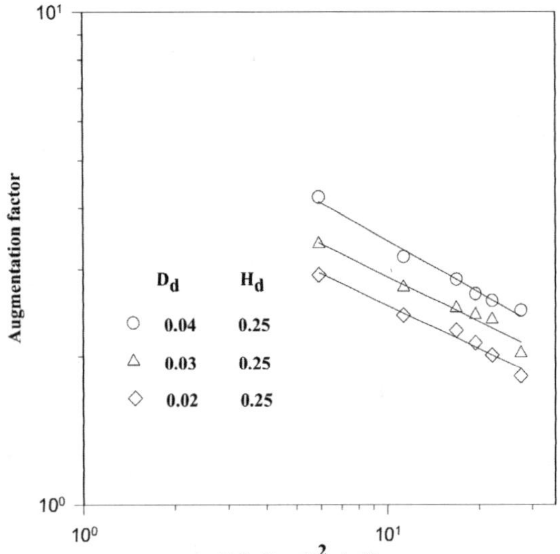

Fig. 10 Variation of augmentation factor with velocity

CONCLUSIONS

1. Mass transfer coefficient is increasing with velocity,

2. Mass transfer coefficient increases with increase in disc diameter.

3. The height of the disc increases to maximum and then showed a decrease. 0.25 m height of the disc was suggested as it presents maximum augmentation.

4. Augmentations achieved, from 1 to 6 folds over smooth tube flow.

5. A model was developed in terms of geometric parameters and presented.

$$\bar{g} = 4593.4\,[Re_m^+]^{-0.00127}\,(\phi_1)^{-0.37357}\,(\phi_2)^{-0.01602}$$

REFERENCES

1. Sherwood, J D and Stone, H A, "Added mass of a disc accelerating within a pipe" *Physics of Fluids -- Nov 1997 – Vol. 9, Iss 11, pp. 3141-3148*
2.. Ravi, T., Srinivasa Rao, B., Gopala Krishna, P and Venkateswarlu, P, "Ionic MT studies in fluidized beds with coaxially placed discs on a rod as internal" *Chemical Engineering and Processing Vol 35, Iss 3, June 1996, pp 187-193.*
3. Burns, J R and Jachuck, R J J, "Deter of liquid solid M Tcoeff for a spinning disc reactor using a limiting current technique" *Int. J of Heat and Mass Transfer Vol 48, Iss 12, June 2005, pp 2540-2547.*
4.Podestá, J J., Paús, G F and Arvía, A J, "Ionic mass transfer at horizontals disc electrodes under longitudinal vibration" *Electrochimica Acta Vol 19, Iss 9, Sep 1974, pp 583-589.*
5. Escudié, R., Epstein, N., Grace, J R and Bi, H T, "Effect of particle shape on liquid-fluidized beds of binary (*and ternary*) solids mixtures" *Chemical Engineering Science Vol 61, Iss 5, March 2006, pp 1528-1539.*
6. Ventateswarulu, P and Jagannadha Raju, G.J.V, *Indian Chemical Engineer Vol. 32, No.13, 101, 1990.*
7.Venkateswarlu, P, "Studies on ionic MT with coaxially placed discs on a rod as turbulence Promoter" *Ph.D., Thesis, Andhra University, Visakhapatnam, INDIA, 1987.*

7. R.G. Deissler, C.S. Eian, "Analytical and experimental investigation of adiabatic turbulent flow in smooth tubes" , *Nat. Adv. Comm. Aeronaut., Washington, DC, TN-2629, 1952.*
8. S.L. Lyons, T.J. Hanratty, J.B. McLaughlin, "Direct numerical simulations of passive heat transfer in turbulen channel flow", *Int. J. Heat and Mass Transfer 34, 1991, pp 1149- 1158.*
9. G.A. Hughmark, "Momentum, heat, and mass transfer analogy for turbulent flow in circular pipes", *Ind. Eng. Chem. Fundamentals 8 , 1969, pp 31-35.*
10. R.G. Deissler, "Analysis of turbulent heat transfer, mass transfer and friction in smooth tubes at high Prandtl and Schmidt". *Nat. Adv. Comm. Aeronaut., Washington, DC, TN-1210, 1955.*
11. E.M. Sparrow, J.A. Malina, "Variable property, constant-property, and entrance-region HT results for turbulent flow of water and oil in a circular tube", *Chem. Eng. Sci. 19, 1964, 953-962.*
12. D.F.Dipprey and R.H.Sabersky, "Heat and momentum transfer in smooth and rough tubes at various Prandtl numbers" *Int J of Heat and Mass Transfer Vol 6, 1963, pp 329-353.*
13. J.Nikuradse, "Laws of flow in rough pipes" *VDI Forsch, 361, 1933.*

NOMENCLATURE

d = diameter of the conduit, m
d_d = disc diameter, m
hd = height of the disc, m
L = length of the conduit, m
g_c = gravity constant, m/s^2
u = local fluid velocity
u^+ =dimensionless velocity
u* = friction velocity
u_{max} = average fluid velocity, m/s
y^+ = dimensionless distance
y_0 = thickness of laminar sub layer
V = velocity of the fluid, m/s
k, C_1, b_1 = constants
Dimensionless
$R(h^+)$ = Roughness momentum transfer function
Rem = Roughness Reynolds number
Re =Reynolds number, $dV\rho/\mu$
f = Friction factor
$\Phi_1 = (d-d_d)/d$
$\Phi_2 = h_d/d$
Greek symbols
τ_0 = wall shear stress, kg/m s^2
ρ = density of the fluid, kg/m^3
μ = viscosity of the fluid, poise
δ = boundary layer thickness
ν = kinematics viscosity
Subscripts
v = viscuous
t = turbulent
b = buffer
I = interface

Table.1 Range of variables covered in the present study

Variable	Minimum	maximum
Disc diameter, m	0.020	0.040
Height of the disc, m	0	0.50
Velocity, m/s	0.0591	0.2751
Reynolds Number, Re	1200	14,000

66d Gas-Liquid Mass Transfer In Horizontal Single Circular Channels Under Slug Flow Regime

Srinivasan Ambatipati and Roshan J J Jachuck. Chemical and Biomolecular Engineering, Process Intensification and Clean Technology (PICT) Group,Clarkson University, PO Box 5707, Potsdam, NY 13699

Gas-liquid mass transfer in circular channels under slug flow regime is a function of diffusivity, channel diameter, average velocity (or bubble velocity), gas and liquid slug lengths[1]. This paper will discuss gas-liquid mass transfer studies currently being carried out by the Process Intensification and Clean Technology Group using horizontal single circular channels under slug flow regime. Volumetric mass transfer coefficient (k_La) values were obtained from a physical desorption process (water-nitrogen) using horizontal circular channels of varying dimensions (1-4 mm) and for a range of flow ratios. High speed camera (500 frames per second) was used to measure the bubble velocity. Data obtained has been compared with a model based on penetration theory proposed by Kreutzer et al[4]. This investigation is expected to establish evidence for the dependence of gas and liquid slug length on mass transfer rates and compliment previous work reported by Bercic et al.,[2] and Vandu et al.[3]

References

1.van Baten, J.M., Krishna, R., 2004. CFD simulations of mass transfer for Taylor bubbles in circular capillaries. Chemical Engineering Science 59, 2535-2545.

2.Bercic, G., Pintar, A., 1997. The role of gas bubbles and liquid slug lengths on mass transport in the Taylor flow through capillaries. Chemical Engineering Science 52, 3709-3719.

3.Vandu, C.O., Liu, H., Krishna, R., 2005. Mass transfer from Taylor bubbles rising in single capillaries. Chemical Engineering Science 60, 6430-6437.

4.Kreutzer, M.T., Kapteijn, F., Moulijn, J.A., Heiszwolf, J.J., 2005. Multiphase monolith reactors: Chemical reaction engineering of segmented flow in microchannels. Chemical Engineering Science 60, 5895-5916.

66e Systematic Analysis Of Process Intensification Options: The Elementary Function Methodology

Hannsjörg Freund, Physical and Chemical Process Engineering, Max Planck Institute for Dynamics of Complex Technical Systems, Sandtorstr. 1, Magdeburg, 39106, Germany and Kai Sundmacher, Process Systems Engineering, Otto-von-Guericke-Univ. Magdeburg, Universitaetsplatz 2, Magdeburg, 39016, Germany.

It is generally agreed that process intensification requires a holistic view of the process, considering the process as a whole system. However, there is no general consensus about the extent of improvements necessary in order to earn the label "process intensification". In our view, these should be drastic improvements of at least an order of magnitude. These improvements may include e.g. a reduction of the plant size by reducing the size of single apparatuses or by reducing the number of the latter by integrating two or more unit operations into a multifunctional unit. Other intensification potentials concern e.g. the specific energy consumption, the amount of reactants used and of waste products being produced for a specified production height (i.e. the feedstock efficiency). A successful intensification of a chemical process requires a systematic debottlenecking which is obtained by identifying and eliminating the main transport resistances that limit the overall process performance and thus can be considered as rate-determining steps on the process level. In our work, we will present an approach that is not based on classical unit operations, but on the analysis of the basic functional principles that are encountered in all processes. Each chemical process can be broken down into five types of elementary function units. Again, it is important to note that we are discussing the elementary function itself (i.e. the underlying physics), and not the function already as part of a unit operation. In the center of the chemical process we define the first elementary function as the "reaction function". The chemical reaction function must be supplied with reactants. In addition, an adequate contacting of the reactants has to be ensured ("contacting function") and the reaction needs to be activated ("activation function"). As a result of the reaction function we generally obtain a mixture of different products (and the fraction of the reactants that have not been converted), so that the introduction of a "separation function" is necessary. Since an energy exchange with the environment takes place during the reaction we finally need an "energy transport function". In our contribution, we will first present the concept of elementary process functions by discussing individually the different functions and the pool of possibilities we have for process intensification measures at each stage. Then we will analyze selected examples of successfully intensified chemical processes and identify and classify the measures according to our elementary function scheme. However, the idea here is not only to optimize an existing process but rather to derive generalized guidelines for the decisions to take for the different elementary process functions in the conceptual design of an intensive chemical process.

94a Developing A Career In Chemical Manufacturing Process Support: A Laboratory Perspective

Hilario Garza, HODer, Shell Global Solutions US, 3333 Highway 6 South, Houston, TX 77082

Providing process support to a chemical plant does not only involve executing routine analytical methods to qualify products or raw materials in a laboratory. It also involves the development and implementation of more elaborate laboratory performance tests and a close interaction between laboratory staff and plant personnel at all levels. This presentation describes the challenges, opportunities and requirements to develop a plant process support career from the viewpoint of laboratory staff.

94b Taking the Bull by the Horns: Developing a Career in Chemical Manufacturing

Jonathan Worstell, HODER, Shell Global Solutions US, 3333 Highway 6 South, Room E-2061, Houston, TX 77082

Most young engineers hired as production engineers at a chemical manufacturing facility are overwhelmed by their day-to-day tasks. Not much time remains in a day for reflective thought about one's career. However, time must be found and thought given to one's career and its development. This talk discusses how to find the time necessary for thinking about one's career and how to plan and develop one's career at a chemical manufacturing facility.

94c What Do You Want To Be When You Grow Up: Career Options And The Skills Needed To Pursue Them

Shawna S. Berg, Geismar Manufacturing Facility, Shell Chemical Company, River Road, Geismar, LA 70734

In our current industrial boom, chemical engineers have a wide degree of options when it comes to careers—far beyond simply being chemical engineers. Over the years, chemical engineers have turned into the "all-inclusive" engineering discipline with jobs ranging from process engineer and development specialists to production managers and sales associates. Some specialty chemical companies will only hire chemical engineers for most areas of business including marketing, strategy, and business managers because of their unique skill set. Chemical engineers are hired for the way they think and solve problems and not simply their knowledge of heat exchangers and reactions. This paper outlines the career options for chemical engineers and the skills necessary to pursue these options.

94d Planning, Selling, Organizing, and Executing a Successful Plant Test

Jonathan Worstell, HODER, Shell Global Solutions US, 3333 Highway 6 South, Room E-2061, Houston, TX 77082

Plant tests must be done to optimize plant operation and to achieve process sustainability. However, plant managers and plant engineers do not consider their production facilities to be pilot plants. Their hesitancy to doing a plant test can be overcome by thoroughly planning and organizing the effort. This talk discusses the planning, selling, and organizing necessary for conducting a plant test. It also discusses what is required for executing a successful plant test.

94e Plant Testing: In-House Or Outside

Mike J. Doll, HODer, Shell Global Solutions, 3333 Highway 6 South, Houston, TX 77082 and Jonathan H. Worstell, HODER, Shell Global Solutions US, 3333 Highway 6 South, Room E-2061, Houston, TX 77082.

It is sometimes not a viable option to conduct plant tests in-house. Reasons for wanting (or needing) to conduct a plant test at an outside facility, i.e. at a contract manufacturer are many, ranging from unavailable capacity to unavailable assets. Conducting a plant test at a contract manufacturer is not unlike conducting a plant test in-house. In this presentation we will compare likenesses and differences in these two options for conducting plant tests.

94f Testing Today's Equipment For Tomorrow: From The Plant Perspective

Jason V. VanRoeyen, Geismar Manufacturing Facility, Shell Chemical Company, River Road, Geismar, LA 70734

Plant tests are a necessary tool to evaluate new production capabilities, optimization, and product portfolio expansions. Despite these potential benefits often times the manufacturing location is hesitant to perform the test from a simple request. That is because with any change from normal operation there are associated risks. Understanding those risks and capturing those risks through a detailed document will build trust amongst stakeholders and increase the probability of a successful plant test.

This paper will explore the type of information that a manufacturing location will look to have answered prior to beginning a plant test. Keep in mind this information exchange not only benefits the manufacturing location, but also those requesting the test to be performed for reasons to be discussed.

151a Process Intensification In Synthesis Of Methyl Acetate

Gregor Demel, Chemical engineering and environmental science, University of Technology Graz, Inffeldgasse 25/c/II, Graz, 8010, Austria, M. Siebenhofer, University of Technology Graz, Department of Chemical Engineering and Environmental Technology, Inffeldgasse 25/C/II, Graz, 8010, Austria, and Rolf Marr, Department of Chemical Engineering and Environmental Technology, Graz University of Technology, Inffeldgasse 25/C/II, Graz, A-8010, Austria.

Processes which combine chemical reaction and product separation in one step are of rising interest in the chemical industry as well as petrochemical and pharmaceutical industry. These reactive systems are so called hybrid processes. Main application of hybrid processes is reactive distillation, a combination of reversible reactions and product separation by distillation. Applicability of reactive distillation is limited to mixtures with different boiling point of substances. Yield of conversion can be increased by evaporating the product from the reaction broth.

Comparable advantages are expected from combination of chemical reaction with several unit operations such as liquid/liquid-extraction. Provided that a solvent can be found which dissolves the product selectively the result of extractive separation of products from reaction mixtures is comparable with distillative product separation except that chemical reaction with extraction can even be operated at ambient temperature. As a consequence lower cost of energy and less complexity of control systems is expected. In this project synthesis of methyl acetate with heterogeneous catalysts has been investigated and modeled. Synthesis was investigated in batch operation and in continuous operation in a plug flow reactor. In both processes esterification was catalyzed with a cation exchange resin Liquid/liquid extraction was carried out with the solvent dodecane. With an appropriate surplus of acetic acid in the feed the formation of the methyl acetate –methanol aceotrop can be avoided because of complete methanol consumption (conversion). Separation of methylacetate from the reaction broth is only affected by the stoichiometric water content. Water/acetic acid-separation by distillation and therefore recovery of the acid surplus does not need a sophisticated rectification set up. The process can be operated without loss of catalyst activity for a very long time (in lab scale it was actually not possible to quantify loss of catalyst activity because it did not obtain a detectable level). By combining esterification with liquid/liquid-extraction the methyl acetate – water aceotrop can be overcome by splitting the product mixture into two separate product fluxes for distillative separation, the binary system acetic acid – water left in the raffinate phase and the binary system methyl acetate – solvent formed in the extract phase.

151b Theoretical Investigation Of Monomer Consumption In Photopolymerization Using Non-Photobleaching Initiator

Venkata Nekkanti, Process Intensification & Clean Technology (PICT) Group, Department of Chemical and Biomolecular Engineering, Clarkson University, Potsdam, NY 13699, Roshan J. J. Jachuck, Process Intensifcation & Clean Technology (PICT) Group, Department of Chemical and Biomolecular Engineering, Clarkson University, PO Box 5705, Potsdam, NY 13699, and Lawrence, M Glasser, for Quantum Device Technology & Department of Physics, Clarkson University, Potsdam, NY 13699.

This paper will discuss a mathematical model developed to predict monomer conversions in a photopolymerization process. The proposed model, which is based on the expression for photobleaching system derived by Wegscheider using Lambert-Beer's law for light absorption in the sample (1), has been validated by experimental data obtained by studying photopolymerization of butyl acrylate and styrene in a narrow channel reactor. Previous mathematical models (2, 3) were based on thin film approximation where the variation of light through the film was neglected. Such a model results in unusually high conversion rates for configurations with high characteristic path lengths.

In establishing our model, experimental data generated by performing a parametric study to investigate the influence of initiator concentration, flow rates, and light intensity, were taken into account. Polymerization was carried out in a narrow channel glass reactor (dia=1.5mm) using UV intensity in the range of 50-150 mW/cm^2. 2,2-dimethoxy-2-phenyl-acetophenone was used as non-photobleaching initiator. Gas Chromatography was used to monitor the conversions and gel permeation chromatography for the molecular weights. This model can also be used for predicting the extent of curing in photolithography applications.

References:

1. Wegscheider, R.; 'Beitrage Zur Photochemischen Kinetik', Z. Phys. Chem. 1923, 103, 273-306.

2. Boodhoo, K. V. K.; Dunk, W. A. E.; Jassim, M. S.; Jachuck, R. J. J.; 'Thin film solvent-free photopolymerization of n-butyl acrylate. I. Static film studies', Journal of Applied Polymer Science, 91 (2004), 2079-2095.

3. Decker, C.; 'The use of UV irradiation in polymerization', Polymer International 45 (1998), 133-141.

151c Intensified Biodiesel Reaction Using Continuous Rotating Tube Reactor Technology

Himanshu Lodha, Chemical Engineering Department, Clarkson University, 8 Clarkson Avenue, P.O. Box 5707, Potsdam, NY 13699 and Roshan J.J. Jachuck, Process Intensification & Clean Technology (PICT) Group, Department of Chemical Engineering, Clarkson University, PO Box: 5705, Potsdam, NY 13699.

This paper will discuss the potential of rotating tube reactor technology for performing transesterification of canola oil using a base catalyst. Rotating tube reactor uses centrifugal force to create thin, intensely mixed liquid films and is capable of delivering high heat and mass transport rates. Using methanol and sodium hydroxide as catalyst, tests have been performed to investigate the influence of rotational speed, flow rate, temperature and catalyst concentration on the reaction rate and the quality of biodiesel in terms of bound glycerol. Atmospheric pressure and low operating temperature (40- 60 degree C) have been used to achieve conversions in excess of 98% in residence times of 40 seconds. In this presentation, explanation of the high rate of reaction will be discussed and the potential for scale up and implications in terms of energy efficiency of biodiesel process plant will be discussed.

151d Fermentation of Glucose into Ethanol by Using An Intensified Reactor

Manognya Billuri, Chemical & Biomolecular Engineering, Clarkson University, 8 clarkson avenue ,PO Box 5707, Potsdam, NY 13699 and Roshan JJ Jachuck, Chemical and Biomolecular Engineering, Clarkson University, 8 clarkson avenue ,PO Box 5707, Potsdam, NY 13699.

This paper will highlight the use of a rotating tube reactor (RTR) for enhancing the rate of fermentation reaction by creating a hydrodynamic environment suitable for biological reactions. Using thin films with controlled shear rates, tests were performed to study the fermentation of glucose into ethanol by using Saccharomyces cerevisiae in a RTR. Rate enhancements by orders of magnitude were obtained in comparison to conventional batch reactor processing. Semi batch operation was performed to demonstrate significant reaction time savings. Thin film concept coupled with the centrifuge action of the RTR is suitable for removal of ethanol from the reacting medium thereby minimizing the inhibition effect of ethanol on the growth mechanism. This inhibition effect is pronounced in the conventional batch reactor operation thereby slowing the speed of fermentation process.

151e Application of Plackett Burman Design and Response Surface Methodology for Optimizing Grinding Kinetics of Dolomite with and without Grinding Aid

Asha Immanuel Raju Chaduvula[1], Murali Yugandhar Nikku[2], Kiran Kumar Nalla[2], Srinivasa Babji Josyula[2], Jagannadha Rao Patruni[2], and Sambi Reddy Bhimavarapu[2]. (1) Chemical Engineering, Andhra University, Andhra University, Visakhapatnam - 530 003, AP, INDIA, Visakhapatnam, 530003, India, (2) Chemical Engineering Department, Andhra University, Andhra University, Visakhapatnam - 530 003, AP, INDIA, Visakhapatnam, 530003, India

ABSTRACT

The assessment of the grinding process quality depends on the amount or quantities of the components used. The present work comprises of optimizing different parameters like time of grinding (t), speed of the ball mill (n), size of balls (Bs), feed quantity (Q), size of feed (Fs) and quantity of additive (Q). The breakage function (k) was correlated with variables of grinding for both the cases of with and without additive. Plackett Burman response surface design was used for determining the regression coefficients and optimum operating parameters for minimizing the power consumption.

Micromixer Characterization with the Consecutive-Competitive Bromination of 1,3,5-Trimethoxybenzene: Experimental Procedure and Comparison to Other Mixing-Sensitive Reactions

Kristin Hecht[1], Andreas Kölbl[2], Manfred Kraut[2], Klaus Schubert[2]
[1]Dept. of Chemical Engineering, The University of Utah, Salt Lake City, UT 84112
[2]Institute for Micro Process Engineering (IMVT), Forschungszentrum Karlsruhe (FZK), Karlsruhe, Germany 76021

Introduction and Background

Mixing is an important process in the chemical industry. Poor mixing can contribute to decreased yields and selectivities. The Handbook of Industrial Mixing estimates that yield losses of 5% are common due to poor mixing[1]. Micromixers offer superior mixing capabilities compared to conventional equipment[2]. Micromixers can mix fluids with a high extent of dispersion and offer shorter lengths of diffusion. Various groups have developed creative devices that mix fluids at the microscale[3-8]. Although each microdevice can offer excellent mixing, it is often difficult to characterize the exact improvement in mixing offered by a micromixer, or to compare small changes in mixer designs. Characterization of the mixing in microdevices is helpful in developing effective devices and comparing them to existing industrial systems and alternative technologies.

Mixing processes have been characterized in numerous ways, both computationally[5,7] and experimentally[9-13]. Computational modeling of mixing is time consuming, and it is difficult to incorporate design complexities, such as the fittings around a microreactor, into simulations. Experimental techniques to gauge mixing include optical phenomena and probe measurements as well as mixing sensitive chemical reactions. Optical phenomena, such as the diffusion of colored or fluorescent dyes, are a quick way to gain a glance at a mixer and develop an intuitive understanding. However, optical phenomena can be difficult to quantify and require an optical window, which is not always readily available. Probe measurements, for example of conductivity or temperature, can disturb the flow, and sufficiently small probes are often not available for microreactors. Mixing sensitive chemical reactions can be applied to complete systems without extensive modifications. The extent of mixing is preserved in the reaction products distribution. However, compared to the other methods, mixing reactions are not able to show the extent of mixing at different locations in a system. They instead can give an overall estimation of mixing in the system in the time before the reactions are complete.

$$A + B \xrightarrow{k_1} R$$

$$C + B \xrightarrow{k_2} S$$

Figure 1: Parallel-Competitive Reaction Scheme

$$A + B \xrightarrow{k_1} R$$

$$R + B \xrightarrow{k_2} S$$

Figure 2: Consecutive-Competitive Reaction Scheme

Mixing sensitive reactions used for characterization are typically composed of two competing reactions. Competitive reaction systems can be comprised of two types: parallel-competitive, shown in Figure 1, and consecutive-competitive, shown in Figure 2. Both reactions should be faster than the mass transport processes to estimate mixing. At the kinetic limit, a certain ratio of the products of the two reactions would exist. Deviations from the kinetic limit can be attributed to insufficient mixing. If k_1 is much larger than k_2, then the molar ratio of S to R should be very small and the concentration of S can be used to quantify the deviation from complete mixing. Two mixing sensitive reactions have been applied for the purpose of comparing micromixing systems in the literature; they are the parallel-competitive reactions of 1-naphthol and 2-naphthol with diazotized sulfanillic acid (azo coupling reaction)[13] and of iodate and dihydrogen borate with acid[9-12]. The reaction of iodate and borate with acid, represented in Figure 3, was developed by the research group of J. Villermaux[14-16] and has been used extensively for microreactor comparisons in literature. It is sometimes called the “Iodide/Iodate Reaction” or “Villermaux Reaction.”

$$5\,I^- + IO_3^- + 6H^+ \xrightarrow{k_1} 6\,H_3BO_3$$

$$6\,H_2BO_3^- + 6H^+ \xrightarrow{k_2} 3\,I_2 + 3\,H_2O$$

Figure 3: Iodide/Iodate Reaction Scheme

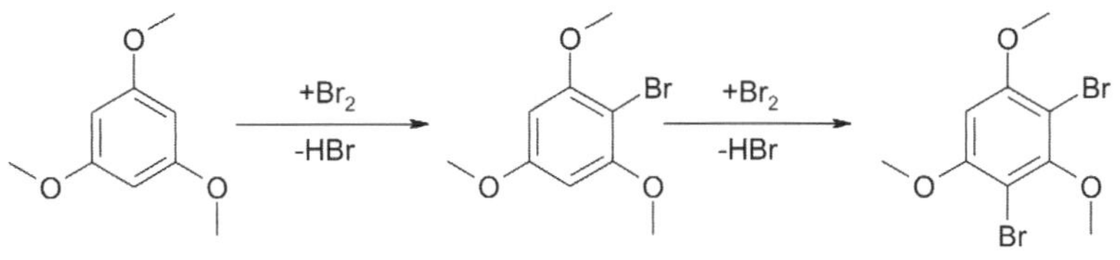

TMB 1BR 2BR

Figure 4: Reaction Scheme for the Bromination of 1,3,5-trimethoxybenzene

In this contribution we discuss the consecutive-competitive bromination of 1,3,5-trimethoxybenzene (TMB) to 2-bromo-1,3,5-trimethoxybenzene (1BR) and 2,4-dibromo-1,3,5-trimethoxybenzene (2BR)[17] and its use in the characterization of microreaction systems. The reaction scheme is presented in Figure 4. We have applied both the Iodide/Iodate Reaction and this Bromination Reaction to mixing devices ranging in size from millimeters to micrometers. Details of application for each method and the results of characterization are presented. The Bromination Reaction Method is compared to the Iodide/Iodate Reaction Method in terms of theoretical advantages, experimental considerations, and the sensitivity and range of mixing characterization.

Experimental Procedure

Three mixing devices were examined with both the Bromination Reaction Method and the Iodide/Iodate Reaction Method. Two T-type mixers with different inner diameters were applied to determine the sensitivity of each characterization method to relatively slight changes in geometry that were expected to only produce slight changes in mixing. **T-Mixer 1**: A union-tee (Swagelok) constructed of nylon with a 4.8 mm inner diameter was used with two axially coincident feed channels and a perpendicular exit. **T-Mixer 2**: A PTFE (polytetrafluoro-ethylene) piece (Bohlender) with a 3.5 mm inner diameter was used in the same manner. **Cyclone-Mixer**: A cyclone-type micromixer fabricated in PTFE at the IMVT was also tested. Details of the cyclone mixer design are shown in Figure 5. Two feeds enter the cyclone mixer and are fed into a swirl chamber in such a way that a three-dimensional rotational flow is created.

A dosing system was established to test each mixer with the two reaction systems. Feed streams were fed at the same mass flow rates. The concentrations of the feed streams used for each method are given in Table 1. For the Iodide/Iodate Reaction Method, the concentrations given in Table 1 were used for characterization of T-Mixer 1 and T-Mixer 2. For the Cyclone-Mixer, triple the concentrations of each compound were used in order to obtain a UV signal. For the Iodide/Iodate Reaction Method, reaction mixtures were examined with UV spectroscopy. The UV absorbance is directly proportional to the concentration of iodine at a

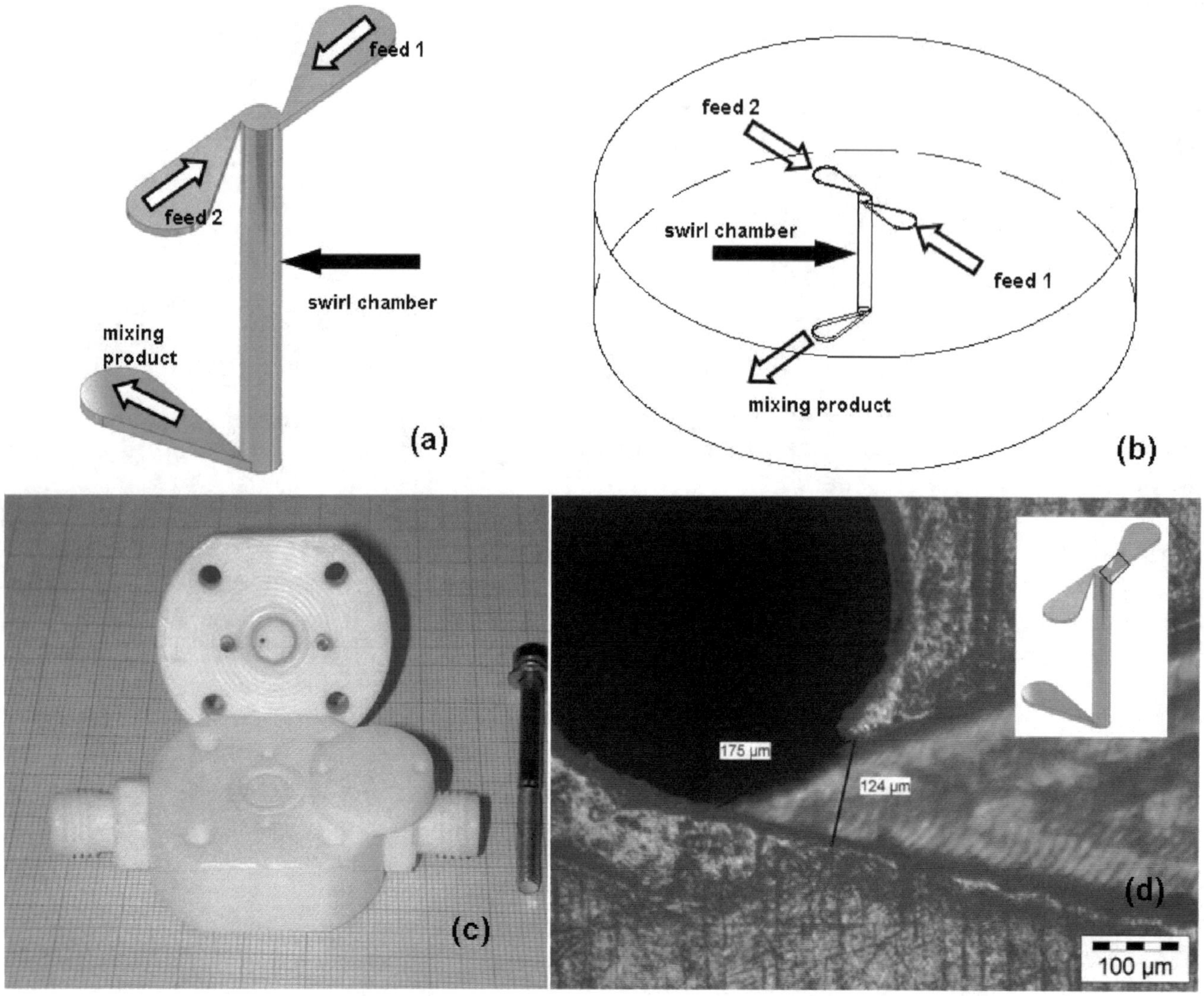

Figure 5: Details of the Cyclone-type Micromixer: (a) Geometry of the cyclone mixer, which is machined into a cylindrical disk to form the micromixer (b) schematic of the disc with mixer geometry included (c) mixer and housing (d) optical micrograph of the edge of one feed entering the swirl chamber

Table 1: Concentrations of the Feed Streams

Iodide/Iodate Reaction Method		**Bromination Reaction Method**	
Feed 1 (water)	**Feed 2 (water)**	**Feed 1 (methanol)**	**Feed 2 (methanol)**
0.04545 M NaOH 0.04545 M H_3BO_3 0.01595 M KI 0.01595 M KIO_3	0.015 M H_2SO_4	0.05 M TMB	0.05 M Bromine

given temperature produced in the reaction[18]. For the Bromination Reaction Method, reaction mixtures were analyzed for the concentration of each reaction component via HPLC[8]. Bromine solutions were freshly prepared prior to each experiment and were covered in aluminum foil during each test to prevent a light-induced reaction of bromine with methanol, which would decrease the concentration of bromine and impact the mass balance[19].

Results and Discussion

The results from testing each mixer using the Bromination Reaction Method are shown in Figure 6. Results are expressed in terms of the concentration of 2BR at a given total flow rate. The concentration of 2BR represents inhomogeneities from incomplete mixing (S in Figure 2). The results for testing T-Mixer 1 and T-Mixer 2 using the Iodide/Iodate Reaction Method are shown in Figure 7; The results for testing the Cyclone-Mixer are shown in Figure 8. Results for the Iodide/Iodate Reaction Method are expressed in terms of UV/VIS absorbance at 351 nm; The UV/VIS absorbance maxima at 351 nm corresponds to the concentration of iodine, which also represents deviations from complete mixing (S in Figure 1).

In each device mixing improves with higher flow rates due to greater energy dissipation. The Bromination Reaction Method was able to characterize each device at the same concentration, which allows for the direct comparison of the devices. With the Iodide/Iodate Reaction Method, the Cyclone-Mixer could not be characterized at the standard concentration because the product concentration was not sufficient to UV/VIS signal. T-Mixer 1 and T-Mixer 2 could not be characterized at triple the concentrations because their absorbance at these concentrations exceeded those that could be measured with the spectrometer. It is not possible to simply dilute samples from the Iodide/Iodate Reaction Method because dilution alters the equilibria occurring between reaction products. The Iodide/Iodate Reaction Method reveals a difference in the mixing performance of T-Mixer 1 and T-Mixer 2 at low flow rates. The differences between these two devices are indistinguishable with the Bromination Reaction Method.

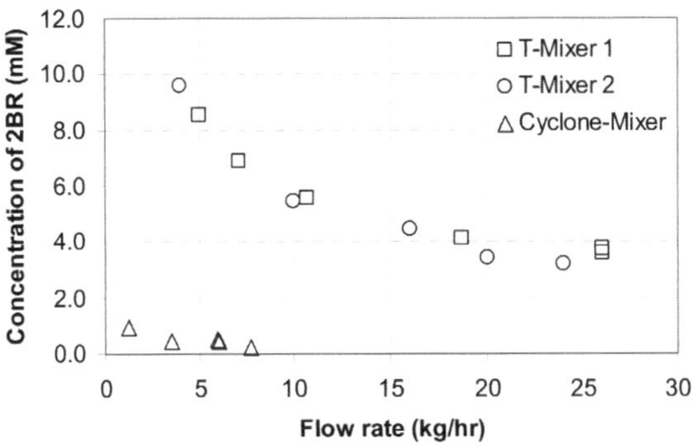

Figure 6: Concentration of 2BR measured for T-Mixer 1, T-Mixer 2, and Cyclone-Mixer using the Bromination Reaction Method

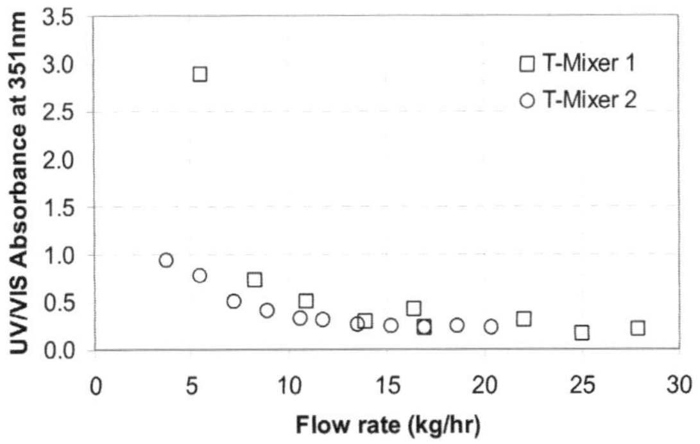

Figure 7: UV/VIS Absorbances obtained for T-Mixer 1 and T-Mixer 2 using the Iodide/Iodate Reaction Method at the standard concentration

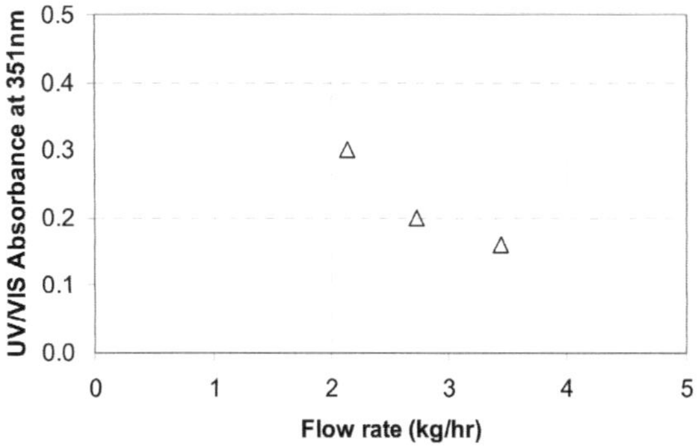

Figure 8: UV/VIS Absorbances obtained for the Cyclone-Mixer using the Iodide/Iodate Reaction Method at triple the standard concentration

Comparison of the Methods

The time to complete mixing measured by these reactions depends on the rate of the reactions, which can depend on reactant concentrations and temperature. In this sense, each set of reactions with a selectivity influenced by mixing could be used for estimating mixing depending on reaction conditions. Both the Iodide/Iodate Reaction Method and Bromination Reaction Method have been shown in this work to be useful for the characterization of mixing in micromixers.

Theoretically, parallel-competitive reaction product distributions can be compared a range of values defined from a concentration of 0 product S (complete mixing is achieved prior to reaction) to a ratio of S to R that is equal to the initial feed concentrations or ratio of A to C (in the case that mass transport is the dominant process and the reaction kinetics are so fast that they do not impact this description of the process.) These limits can be mathematically convenient and are used to determine a value that is sometimes called the "segregation index." The segregation index is not the same as the extent of mixing or homogeneity, and it is only accurate to compare the segregation index for two systems when they are characterized with the same reactions and the same reactant concentrations. Such an index cannot be easily assigned to consecutive-competitive reactions.

Experimentally, both reaction characterization methods face some limitations. Bromine is not compatible with steel and must be used with compatible materials such as glass, most plastics, and compatible metals such as gold. Iodine may absorb onto surfaces under some conditions. The HPLC analytical method employed by the Bromination Reaction Method allows for the construction of a mass balance, but the UV/VIS method used by the Iodide/Iodate Reaction Method can only be used to obtain the concentration of one species.

The Iodide/Iodate Reaction Method was able to show slight differences in the mixing qualities between T-Mixer 1 and T-Mixer 2, which differed only slightly in design. The Bromination Reaction Method was not able to distinguish between the slight differences in mixing in the two devices.

The Bromination Reaction Method was able to characterize each device at the same concentration. Different concentrations were required for the devices with the Iodide/Iodate Reaction Method. It is convenient to have a method that uses the same reactant concentrations for different devices because the devices can be compared directly and comparison among different authors using the same concentrations would be possible.

References

1. Handbook of Industrial Mixing: Science and Practice. Paul E.L.; Atiemo-Obeng V.A.; Kresta S.M., Eds. Hoboken, New Jersey: John Wiley & Sons, Inc., 2004.
2. Hessel V.; Hardt S.; Löwe H. Chemical Micro Process Engineering: Fundamentals, Modelling and Reactions. Weinheim, Germany: Wiley-VCH Verlag GmbH & Co., 2004.
3. Bertsch A.; Heimgartner S.; Cousseau P.; Renoud P. Static Micromixers Based on Large-Scale Industrial Mixer Geometry. Lab on a Chip. 2001, 1, 56-60.
4. Liu R.H.; Stremler M.A.; Sharp K.V.; Olsen M.G.; Santiago J.G.; Adrian R.J.; Aref H.; Beebe D.J. Passive Mixing in a Three-Dimensional Serpentine Microchannel. Journal of Microelectromechanical Systems. 2000, 9(2), 190-197.
5. Mengeaud V.; Josserand J.; Girault H.H. Mixing Processes in a Zigzag Microchannel: Finite Element Simulations and Optical Study. Analytical Chemistry. 2002, 74, 4279-4286.

6. Stroock A.D.; Dertinger S.K.; Whitesides G.M.; Ajdari A. Patterning Flows Using Grooved Surfaces. Analytical Chemistry, 2002, 74, 5306-5312.
7. Wang H., Iovenitti P., Harvey E., Masood S. Optimizing Layout of Obstacles for Enhanced Mixing in Microchannels. Smart Materials and Structures, 2002, 11, 662-667.
8. Hecht K., Kraut M., Kölbl A. Microstructured Mixing Devices: An Efficient Tool for the Determination of Chemical Kinetic Data? AIChE Spring Meeting, Houston TX, April 2007.
9. Paníc S.; Loebbecke S.; Tuercke T.; Antes J.; Bošković D. Experimental Approaches to a Better Understanding of Mixing Performance of Microfluidic Devices. Chem. Eng. J. 2004, 101, 409-419.
10. Nagasawa H.; Aoki N.; Mae K. Design of a New Micromixer for Instant Mixing Based on the Collision of Micro Segments. Chem. Eng. Technol. 2005, 28, 324-330.
11. Ehrfeld W.; Golbig K.; Hessel V.; Löwe H.; Richter T. Characterization of Mixing in Micromixers by a Test Reaction: Single Mixing Units and Mixer Arrays. Ind. Eng. Chem. Res. 1999, 38, 1075-1082.
12. Werner B.; Hessel V.; Löb P. Mixers with Microstructured Foils for Chemical Production Purposes. Chem. Eng. Technol. 2005, 28, 401-407.
13. Ehlers S.; Elgeti K.; Menzel T.; Wießmeier G. Mixing in the Offstream of a Microchannel System. Chem. Eng. Process. 2000, 39, 291-298.
14. Fournier M.-C.; Falk L.; Villermaux J. A New Parallel-competitive Reaction System for Assessing Micromixing Efficiency—Experimental Approach. Chem. Eng. Sci. 1996, 22, 5053-5064.
15. Villermaux J.; Falk L.; Fournier M.-C.; Detrez C. Use of Parallel-competitive Reactions to Characterize Micromixing Efficiency. AIChE Symp. Ser. 1992, 88, 6-10.
16. Villermaux J.; Falk L.; Fournier M.-C. Potential Use of a New Parallel Reaction System to Characterize Micromixing in Stirred Reactors. AIChE Symp. Ser. 1994, 90, 50-54.
17. Bourne J.R.; Kozicki F. Mixing Effects During the Bromination of 1,3,5-trimethoxybenzene. Chem. Eng. Sci. 1977, 32, 1538-1539.
18. Awdrey A.D.; Connick R.E. The Absorption Spectra of I_2, I_3^-, I^-, IO_3^-, $S_4O_6^=$ and $S_2O_3^=$. Heat of the Reaction $I_3^- = I_2 + I^-$. Journal of the American Chemical Society. 1951, 13, 673-683.
19. Hecht K.J. Characterization of Mixing Processes Through the Consecutive-Competitive Bromination of 1,3,5-Trimethoxybenzene. Thesis. University of Utah. 2007.

245a New Product Development And Related Topics In Undergraduate Engineering

Nick Nissing, Energy, Environmental, and Chemical Engineering, Washington University, One Brookings Drive, Box 1180, St. Louis, MO 63130

New Product Development has become an integral part of the chemical engineering curriculum at Washington University, through the addition of two new courses, one of which is required for the undergraduate degree. These courses have created an excellent opportunity to directly address topics like intellectual property, creativity, entrepreneurship, and open-ended problem solving, while working in teams. In this context, a recent survey of academic perceptions related to teaching new product development, entrepreneurship, and intellectual property was conducted. 102 responses were received from engineering school deans, ChE department chairs, and professors. Conclusions include observations regarding academic priorities related to these topics, and key barriers to changing current practice.

245b Simultaneous Product And Process Design

Aydin Sunol, Department of Chemical Engineering, University of South Florida, 4202 East Fowler Avenue, ENB 118, Tampa, FL 33620

This presentation is based on our experience in transformation of Design curriculum over the past two decades at institutions the author have been affiliated with and focus on the challenges associated with integrating various elements of product and process design within the projects and curriculum. Projects from diverse emerging areas in microelectronics, fuels cells, targeted delivery products, biomedical devices, and novel material are used. The utility of core processing technologies in the design of several different products will be illustrated using products such as self heating/cooling products, sustainable energy based products such as green sail boats, and targetted delivery products. The impact of the product design on the curriculum including the efforts in recent NSF Spiral Curriculum Reform Implementation grant at the University of South Florida will also be discussed. The selected projects involve novel products that has been introduced to market by our industrial partners. Uncertainty analysis, microeconomics, and Computational Fluid Dynamics are heavily relied on in the projects.

245c Teaching Product Design Through Integrated, Multiscale Projects

Kenneth R. Cox, Rice University, 6100 Main Street, Houston, TX 77005-1892

We are confronted with the great challenge of incorporating product design concepts into a curriculum that already constrained by limits on total hours and total numbers of courses. While one approach would split product design and process design into separate tracks, a promising alternative involves wrapping both product and process design into a single capstone project. The latter treatment builds on common features of both scales of design, while emphasizing optimization opportunities built on the integrated, multiscale perspective.

Process design projects in reality include much more than the design of tubes and tanks. All processes offer improvement opportunities based on a knowledge of molecular morphologies (such as crystallization), the use of nanostructuring compounds (such as corrosion inhibitors), leveraging mesoscale structures (such as membranes), or employing advanced macroscale structures (such as structured packing). Each of these are examples of the opportunities presented in process design projects.

Through the analysis of separations processes, we demonstrate how we can use this understanding to seamlessly incorporate product design studies into existing curricula.

245d Teaching Product Design Using Microeconomics

Miguel J. Bagajewicz, Chemical, Biological and Materials Engineering, University of Oklahoma, 100 E. Boyd St., T-335, Norman, OK 73019

At the University of Oklahoma, we teach product design in the capstone class. To do that, we identify different interesting products. Some examples are: lotions, insect repellents, wine, carpet deodorizers, oxygen generators, artificial nose, medical diagnostic devices, vaccines, beer, etc.

Many of these project results were presented in AIChE meetings as oral presentations. They use a methodology rooted in connecting product properties (compositions, structure, functionalities) with consumer preferences and production costs. In addition, we use pricing models to assess what is the demand price (hedonic) relationship.

Students learn to build models in which product properties are manipulated to maximize consumer preference or to maximize profit. Both give different results.

WE also touch on competition, uncertainty and financial risk considerations.

Illustrations will be made.

245e Stochastic Financial Analysis in Product Design

Michael Hill, M. Hill & Associates, 31 Winding Trail, Mahwah, NJ 07430

Product and process design problems generally require utilization of parameters that are not known with absolute certainty, whether they are physical property parameters, chemical conversions and selectivities, anticipated costs of raw materials, or other design criteria. While most chemical engineering textbooks assume financial analysis is deterministic, stochastic methods are more appropriate and easy to apply.

The capstone course in Chemical Product Design at Columbia University has been teaching undergraduate students to apply probability theory when performing financial analyses. The instruction has included guidance for choosing appropriate probability density functions, and two methodologies for performing calculations with probability density functions, namely the method of moments and Monte Carlo simulation. Students have also been given guidelines as to when each technique is more appropriate. As a result, students are able to calculate financial decision criteria such as breakeven sales volume and net present value to a specified confidence level.

While these techniques have been taught in a context of product design, they are equally appropriate for process design.

309a Multiscale Modeling - An Industrial Perspective

Lionel O'Young, ClearWaterBay Technology, 20311 Valley Blvd, suite C, Walnut, CA 91789

Text Not Available.

Engineering Skin Lotion

Miguel Bagajewicz, Season Hill, Amanda Robben, Heyde Lopez, Monica Sanders, Erin Sposato, Curtis Baade and Shamara Manora

School of Chemical, Biological and Materials Engineering
University of Oklahoma, Norman, OK 73019

Introduction

A lotion was engineered to help people diagnosed with Ichthyosis Vulgaris. People who have Ichthyosis Vulgaris (a genetic disorder that causes extreme scaly, dry skin) want to buy a lotion that does more than replenish the water in the skin for an hour or two. The lotion needs to help with the desquamation process as well; a process that involves the shedding of dead skin cells.

A model has been developed that is able to choose the lotion's ingredients and relative concentrations based on consumer preferences and a pricing model. The consumer preferences were found by summing the product of the importance of a lotions attribute (e.g. effectiveness, smoothness, thickness, etc.) to a consumer's satisfaction. The lotion's attributes were related to a physical property of the lotion (e.g. viscosity, surface tension, etc.). These relationships were used to optimize the lotion's contents.

A demand model was also created to find the most profitable product, by varying the ingredients and concentrations of the lotion. The perfect product can be produced non-profitably with some compromise. Microeconomic variables that vary the profitability of the lotion include demand, price, costs, advertising and capital investment.

Background on Skin

In order to understand what a lotion needs to do, one must first understand the layers of the skin and what happens to skin when it is dry. Moisturizing effects take place in the outer most layer of the skin, the epidermis. Moreover, the stratum corneum (SC) is the outer most layer of the epidermis. The water content in the SC is important for keeping the desquamation process active and the skin an effective barrier against infection. The SC also contains natural moisturizing factors to help absorb and retain water in the skin.

Background on Skin Disorders

Ichthyosis is a family of genetically inherited disorders characterized by severely scaly skin. A genetic defect leads to a buildup of the SC which produces the dry scales on the skin. Since Ichthyosis is a genetic disorder, there is no cure and its treatment centers on managing the symptoms. The most common form of Ichthyosis is Ichthyosis Vulgaris, accounting for 95% of all Ichthyosis cases. The symptomatic treatment of Ichthyosis Vulgaris involves using moisturizing lotions to restore the lipid bilayer of the SC, deliver moisturizing agents to the skin, and promote desquamation.

Lotions

The different types of ingredients used in lotion can be separated into active and inactive ingredients. Active ingredients are the chemicals that help with the function of the lotion. The active ingredients help attract and bind water to the skin, prevent loss of water from the skin, promote dead skin removal, and fill the intercellular spaces on surface of skin. The inactive ingredients do not affect the function of the lotion, but they make it possible to use the active ingredients. The inactive ingredients have a variety of functions including stabilizing the emulsion, preventing contamination of the lotion by microbial organisms, and maintaining the viscosity and pH of the lotion.

Lotions are oil in water emulsions. The solvent is water while some of the ingredients are insoluble in water. Emulsifying agents are used to stabilize the oil in water solution. The actual manufacturing procedure consists of mixing the oil and water phases together.

Consumer Preference Model

In order to quantify consumer preference, a model was developed based on the ingredients in the lotion. The consumer preference score of product i (P_i) was defined as a weighted average of normalized scores of different consumer related "properties" ($y_{i,j}$).

$$P_i = \sum y_{i,j} w_i \qquad (1)$$

where w_i are the weights of importance for a particular attribute. The scores were defined in the range from zero to one. The attributes of the lotion was related to an adjustable physical property. These relationships are shown in Table 1.

Table 1. Lotion Attributes Relationship to Physical Properties

Consumer Properties	Physical Properties
Effectiveness	Diffusion
Ease of Application	Surface Tension
Thickness	Viscosity
Smoothness	Coefficient of Friction
Creaminess	Viscosity & Concentration of Insolubles
Absorption Rate	Diffusion Steady State Time
Greasiness	Concentration of Insolubles

Each consumer property was critiqued on a scale from 0-10, going from one extreme to the next (i.e. very thin to very thick, not greasy to very greasy, etc.) from a small scale informal market analysis. These scores were graphed verse the consumer preference. Moreover, the consumer rating was also graphically compared to a physical property.

Maximizing Consumer Preference

Microsoft Excel was used to calculate the ($y_{i,j}$) value for each property. The weights of importance were determined for the Northwest, Northeast, Southwest, and Southeast regions of the United States by studying the climate's affect on those with Ichthyosis Vulgaris. The weight of importance also varied based on age and gender. Solver in Excel was set up to change the weights as well as the concentrations of the ingredients to maximize the consumer preference. Solver found that the southwest region is the best region to market the lotion.

In addition to choosing the concentrations of the ingredients, Solver also chose which ingredients to include in the lotion. A constraint was added that the sum of the concentration for each type of ingredient needed to be within the following bounds. Using all of these constraints, Solver found the max preference to be 84%. While this lotion provides the largest preference for consumers, it is not economically feasible to produce.

In order to make the process economically feasible, a demand model was created. The demand model changed the ingredients and the concentrations used to find the most economical solution. The product price and product demand was also optimized.

Price-Demand-Quality Model

The demand model was based on the following equation.

$$\beta p_1 d_1 = \alpha p_2 d_2 \tag{2}$$

where, $\beta = \frac{H_2}{H_1} = \frac{Competitors Happiness}{Our Products Happiness}$

α=awareness of existence of new product

p_1=our price

p_2=competitors price

d_1=our demand

d_2=competitors demand

With the total demand (D) equal to d_1+d_2, the demand equation can be rearranged to find our demand.

$$d_1 = \frac{D}{\left(1 + \frac{\beta \cdot p_1}{\alpha \cdot p_2}\right)} \tag{3}$$

Using this equation, our demand was found by solver with the constraint that d_1 needed to be less than or equal to equation (3). The overall demand was found by multiplying the population of the market region by 15.4%. This percentage came from 0.4% of the population having ichthyosis vulgaris and assuming that 15% of the population has xerosis at least once each year. From this number it was assumed that only 10% actually purchase lotion for their symptoms. For our maximum preference, it was assumed that since it would be the highest quality product on the market, that our product would have a saturated market. Therefore, equation (3) was used without changing to an inequality.

To find the most profitable lotion, the return on investment (ROI) was maximized. The ROI was calculated using the following equation.[1]

$$ROI = \frac{CF}{TCI} = \frac{(s - c_o - d)(1 - \phi) + d}{TCI} \tag{4}$$

where, s=sales=$\Sigma p_1 d_1$

c_o=cost=$\Sigma x_i d_1$-administration-advertising-transportaion-packaging

φ=0.35

α was assumed to be 1. Alpha assumes the population knows about our product and the competitor's product the same amount.

Business Model Results

In order to find β, the competitor's preference needed to be found using our preference models. Using the Cosmetic & Toiletry Formulation Book, the ingredients and concentrations for one of the competitors was found.[2] A preference of 51% was found for the competitor when the competitor's ingredients and concentrations were plugged into the preference model. Using the business model, the preference of a most profitable lotion product was found to be 71%.

Conclusions

The preference model showed that a perfect product is not economically feasible. The highest preference was found to be 84%, with a return on investment (ROI) of $-1.44*10^4$%. A demand model was then created to find the most profitable solution. The demand model maximized the ROI, while changing the demand, lotion ingredients and ingredient concentration.

From the demand model, the most profitable product had a preference of 71% with an ROI of 40%. The competitor's preference is 51% based on the preference model. A risk analysis showed that the demand has the greatest effect on the ROI. As long as the demand is high, the process is profitable.

While the models give a good approximation on the lotion product and process, there are still flaws to be worked out. There are a few ingredients that are in the competitor's lotion, but not in the preference model. In order to have a better comparison between our product and the competitors, the model needs to have all of the competitor's ingredients to choose from. Microsoft Excel's Solver was also found to not have the capabilities to completely handle all of the constraints. Some of the calculations had to be manipulated manually instead of Solver going through all of the options. Another software program should be looked into that can better handle then number of constraints and the non-linear equations.

Acknowledgements. Thanks to Hermia, Inc. who helped with the background research.

References

[1] Peters, Timmerhaus, & West. (2003). *Plant Design and Economics for Chemical Engineers.* (5th ed.). Boston: McGraw-Hill.

[2] Flick, E.W. (2001). *Cosmetic and Toiletry Formulations.* (2nd ed.). Noyes: William Andrew Publishing. Retrieved February 10th, 2006, from Knovel Library website: http://knovel.com/knovel2/Toc.jsp

309c Examples of Multiscale Modelling in Chemical Product Design

Rafiqul Gani, Department of Chemical Engineering, Technical University of Denmark, Building 229, Søltoft Plads, Kgs. Lyngby, Dk-2800, Denmark

One alternative in design of structured chemical products is to identify the desired end-use properties of the product and then to match this value by manipulation (design) of the product at the microstructure level. For development and/or manufacture of the same structured product, the end-use properties of the product are controlled by manipulating the microstructure formation of the final product. Typical examples are polymeric structure-based chemical products where the end-use of the polymer defines the product function (or performance) while the properties of its microstructure defines its design and manufacture. Clearly, any model-based design of these structured products will require models at micro-scale (related to properties of the microstructure) as well as macro-scale (related to the end-use properties of the final product). In this case, the macro-scale (end-use) property of the product could be regarded as an average property of its microstructure, calculated through an appropriate population balance model. The important parameters in the population balance model could be, for example, a combination of number of microstructures, their size and their distribution, or, a set of molecular (or atomic) descriptors. That is, the population balance model parameters are the links between the models at different scales of size. An interesting feature of the design of structured chemical products is that while the property of the product at the macro-scale and at micro-scale may be unique, the way to link them is not unique. That is, different macro-scale structures with the same end-use property maybe obtained through different arrangements of the same micro-scale. Alternatively, different microstructures arranged in different forms could be made to give the same end-use function. A large number of chemical products from the pharmaceutical, food, agro-chemicals, cosmetics, etc., sectors show the above "micro-macro" relationship. While development of these products has been and continue to be experiment based trial and error, use of modelling tools, specially those with multiscale modelling features are finding increasing use. Multiscale modelling combined with experiments, has the potential to significantly reduce the time and cost to market for these structured chemical products. The presentation will highlight some of the important issues related to the use of multiscale modelling in structured chemical product design through illustrative examples. For each example, the representation of the structure at different scales, the definition of properties, the linking parameters, the multiscale model features, the simulation strategy, and, the chemical product design problem formulation and solution, will be highlighted.

309d Multiscale Modeling for Polymer Nano-Composite Coating Development

Jie Xiao and **Yinlun Huang**. Department of Chemical Engineering and Materials Science, Wayne State University, 5050 Anthony Wayne Drive, Detroit, MI 48202

Polymer nano-composite coating (or simply called nano-coating) has been of great interest, as it may offer new or enhanced coating performances. Current nano-coating development approach has been exclusively experiment based. In laboratory, a small amount of organo-modified nano-particles is introduced into a conventional waterborne or solvent-borne resin. The resulting resin is then applied on a small substrate surface and a layer of thin film is formed through curing under a set of well-controlled curing conditions. A desired coating formulation is expected to be identified through repeated experiments. This approach does not support comprehensive and extensive experiments due to the concern of experimental cost and time limit. Furthermore, the experimental conditions are very different from the real coating manufacturing conditions. This could cause the nano-coating properties designed in lab to be not fully realizable in the final coating in real application.

In this work, a multiscale modeling approach is applied for effective nano-coating development in the automotive industry. By this approach, a set of multiscale models are developed to establish comprehensive correlations among material properties (at nano-to-macroscale), application conditions (usually at macroscale), and coating performances (at meso-to-macroscale). The macroscopic oven curing model set includes an air flow turbulence model, a radiation heating model and a panel heating model using CFD. This model can generate location specific coating temperature information. A lattice Monte Carlo with bond fluctuation model is utilized to construct the coating nano-to-mesoscale structure, i.e., the connectivity in the polymer network, spatial distribution of nanoparticles, and polymer chain conformation. A coating quality model correlates coating structure with the final coating property (e.g., stiffness). The information passed from the macroscopic curing condition model to the Monte Carlo code is generated through a macro-micro integration method.

The effects of material parameters and curing conditions on the final coating stiffness are thoroughly investigated. Simulation has revealed that when the nanoparticle and polymer have attractive interactions, the coating stiffness is steadily increased as increasing nanoparticle concentration. It is also suggested by simulation that the coating quality uniformity throughout the vehicle body can be improved through changing curing conditions (i.e., the direction of the air sprayed from the nozzles). These insights are especially valuable for identifying improved material formulation and application conditions, so that experiment based nano-coating development can be greatly assisted and product development cycle time can be significantly reduced.

309e Integrating Mixture Design within the Property Clustering Framework

Charles C. Solvason, Nishanth Chemmangattuvalappil, and Mario R. Eden. Department of Chemical Engineering, Auburn University, 132 Ross Hall, Auburn University, AL 36849-5127

Mixture Design is a Design of Experiments (DOE) tool used to determine the optimum combination of chemical constituents that deliver a desired response (or property) using a minimum number of experimental runs. While the approach is sufficient for most experimental designs, it suffers from combinatorial explosion when dealing with the multi-component mixtures found in pharmaceutical excipient design. To circumvent this problem, a recently developed design technique called Property Clustering is applied. In this type of design the properties are transformed to conserved surrogate property clusters described by property operators, which have linear mixing rules even if the operators themselves are nonlinear. Product and process property targets are then used to describe a feasibility target region. To solve the mixture design, components are mixed according to property operator models in a reverse problem format until the mixture falls within the feasibility target region. Once candidate solutions are found, they can be screened with additional criteria per the experimenter's preference.

The degree of accuracy of this modeling technique depends heavily on the ability of the property operator models to adequately describe the property within the studied design space. This work focuses on the utilization of linear, or first order, Scheffe canonical and Cox polynomial models as the property operators. It is shown that the Scheffe models can be used in the existing clustering format, while Cox models require the clusters to be rewritten in terms of standard and a pseudo clusters. While both models result in identical solutions, it is shown that use of the Cox model reduces collinearity between the chemical constituents and leads to a set of interpretation opportunities unrealized when using Scheffe models. It is also shown that the use of statistically derived property operators may lead to negative property cluster space for which techniques are presented on how to quantify and qualify such regions.

In situations of high collinearity, property clustering is applied to the latent variables associated with Principal Component Regression (PCR). It is shown that representing the principal latent properties in the cluster domain not only reduces the number of plots required to illustrate the latent variable space, but also provides a medium from which components and component mixtures can be selected for experimental design. In situations where a component or component mixture does not meet the selection criteria, new molecules can be created based on group contribution methods. In addition, it is shown that process constraints can be mapped into the latent variable space to facilitate a more efficient design.

Selected case studies in mixture and molecular design will be used to illustrate the techniques.

375a Designing Wine Under Uncertainty

Miguel J. Bagajewicz, Ashley Heller, and Craig Whitnack. Chemical, Biological and Materials Engineering, University of Oklahoma, 100 E. Boyd St., T-335, Norman, OK 73019

An existing methodology for the development of consumer products, one that builds consumer preference functions and uses pricing models, is applied to winemaking. Consumer needs and preferences are identified, allowing the data generated by market analysis to be related to wine properties. These wine properties are easily engineered throughout the winemaking process and can be manipulated by the manufacturer at little cost. This new consumer function can be incorporated into a demand model that allows for the manipulation of the selling price of a bottle of wine. Based on the consumer and the demand models, maximization of the profit can be performed. Although we understand that wine can be described by many more characteristics, we picked the following: Clarity, Color, Bouquet, Acidity, Sweetness, Bitterness, Body/Texture, Finish/Aftertaste. We then connected these characteristics to specific wine properties and processes, especially to the compounds that are obtained through the different levels of oak barrel toasting. The summation of the products of each attribute and their corresponding weights allow for the development of a preference function. The value of satisfaction of the consumer is then compared to that of the competition, forming the superiority function that governs the demand model. Without uncertainty incorporated into the demand or business model, the optimum production capacity of wine was found. When uncertainty is incorporated into the model, the decision making changes and the risk associated is too large. It was found that manufacturing a wine of lower consumer utility can, in some cases, reduce the risk at similar aspiration levels.

375b Engineering Roach Killers: Population Models, Pheromone Manufacturing and Consumer Pricing Models

Miguel J. Bagajewicz[1], Doug Beshara[2], and Anthony Williams[2]. (1) Chemical, Biological and Materials Engineering, University of Oklahoma, 100 E. Boyd St., T-335, Norman, OK 73019, (2) CBME, University of Oklahoma, 100 E. Boyd St., T-335, Norman, OK 73019

We address the possibility of integrating a synthetic roach sex pheromone into an existing trapping mechanism to produce a superior roach control product. This new product would be an off the shelf brand designed for private homeowners, and should be competitive with the current market leader product. It was decided to market this product in a gel form that combines blattellaquinone (the pheromone) and maltose as the attractants, fipronil as the active insecticide, and carrageenan as the gel matrix. The southwestern United States (TX, OK, KS, NV, AZ, NM, CA, UT, CO) is the target distribution area. Utility functions were created to assess the necessary characteristics of the new product from a consumer standpoint. Four characteristics of the new product were rated by consumers through the distribution of surveys. The four characteristics were: durability (how long before roaches come back), speed (how fast are roaches killd), odor, and toxicity. Each of these characteristics were then linked to an adjustable component of the new product such that shifting the composition of the product altered consumer satisfaction ratings accordingly. The resulting utility functions were then used in combination with an economics based pricing model correlation to analyze how product composition affects supply and demand, plant size, FCI, and ultimately, profit. To adress durability and speed a population model was constructed. To address the fabrication of the pheromone, plant design was performed so that the economics of the FCI and the operating costs can be assessed.

375c Slow Release Carpet Deodorizers/disinfectants- An Exercise in Mass Transfer, Microeconomics and Uncertainty Management

Miguel J. Bagajewicz[1], Carrie Street[2], Justin Woody[2], and Jaime Ardila[2]. (1) Chemical, Biological and Materials Engineering, University of Oklahoma, 100 E. Boyd St., T-335, Norman, OK 73019, (2) CBME, University of Oklahoma, 100 E. Boyd St., T-335, Norman, OK 73019

Carpet can serve as a haven for dust mites, mold, mildew, and bacteria. A freshening and disinfecting powder product was designed to address these problems and provide a long lasting fragrance. The product includes baking soda to absorb moisture, boric acid to serve as a disinfectant, and linalool (lily) fragrance in PLGA (polymer) particles to provide a long lasting scent.

A consumer satisfaction and pricing model was used to determine the best particle sizes and fragrance concentrations, as well as proportions of all ingredients.

The following consumer utilities were considered in the design of this product: disinfectant effectiveness, scent intensity, fragrance duration, toxicity, odor elimination, and scent type.

Disinfectant effectiveness (related to boric acid contents) is measured by the percentage of mites killed per unit area of carpet. A mass transfer model allowed relating the fragrance particle dimensions to the scent intensity at a plane five feet above the carpet, and to the fragrance duration. The toxicity and odor elimination utilities allowed for the determination of the amount of boric acid and baking soda per unit area of carpet, respectively.

375d Assessment of Performance Risk in Product Design & Development

Michael Hill, M. Hill & Associates, 31 Winding Trail, Mahwah, NJ 07430, **Lynn H. Lander**, Lander & Associates, LLC, PO Box 201, Harrington Park, NJ 07640-0201, and David Parrott, Unilever R&D Port Sunlight, Quarry Road East, Bebington, Wirral, United Kingdom.

Product design and development projects are subject to substantial uncertainties termed "project risks". Unless these risks are identified and managed prior to product launch, they can lead, not only to project failure, but can even create dire consequences for the entire business enterprise.

While conventional project risks often relate to time and cost issues which readily lend themselves to quantitative treatment, risks in product design and development projects more typically relate to performance issues where the knowledge base is inadequate for quantitative treatment. For example, performace issues may include uncertainties related to the market's need for the product being designed, the technical feasibility of meeting the design requirements, the adequacy of the skills of the design team, the supply chain capability of the business enterprise, and the response of competitors in the marketplace, to name just a few. Although these issues may be quantified through the assignment of arbitrary numeric scales, they are semi-quantitative at best and hence require non-quantitative techniques to identify, assess and manage them.

This paper will present a methodology developed by a major multinational product-centered company to analyze performance risk in product design and development projects. It uses a checklist of some 25 issues that all product design and development projects must consider during project planning and execution. These issues are assessed by project teams using a system of "anchored scales" to reduce misunderstanding and subjectivity and thereby aid consistency of judgement. The output is a risk plan that becomes an integrated part of the project plan, including metrics for determining whether the identified risks are being mitigated over time and signals to monitor for negative events.

Based on over a decade's experience, this technique has been proven to be a useful tool for helping teams cope with the uncertainties of new product design and development.

421a High Throughput Experimentation to Support Process Optimization

Eric Carlson, Symyx Technolgies Inc, 415 Oakmead Parkway, Sunnyvale, CA 94085

In process development labs, there is a need to broadly profile compound solubility, to perform crystallization and salt selection studies, and perform process optimization studies. The use and impact of automated instrumentation and integrated workflows to perform these studies will be discussed.

421b Solubility Modeling from High Throughput Solvent Screening

Jose E. Tabora, Research and Development, Bristol-Myers Squibb, 1 Squibb Dr., New Brunswick, NJ 08901 and Chau-Chyun Chen, Aspen Technology, Inc., 10 Canal Park, Cambridge, MA 02141.

High throughput techniques have found widespread application in pharmaceutical research and development. In the design of crystallization in particular, measuring solubility in a few dozen solvents is a commonly performed screening procedure to identify optimal solvent systems. However, due to the enormous potential combination of binary and ternary systems that may be employed, a purely empirical approach to optimize the crystallization solvent composition is intractable. Solubility modeling techniques are typically used to aid crystallization design by estimating the solubility of interest in untested systems. NRTL-SAC has proved a powerful tool to estimate solubility by modeling the activity coefficient of the solute with four adjustable parameters which are fitted from experimental data. Typically five to ten experimental values are sufficient to obtain the universal solute descriptors of the NRTL-SAC model. We present in this paper a systematic approach to optimize the data set used to fit and validate the NRTL-SAC parameters from experimental solubility obtained with high throughput technology.

421c Intensifying Process Development Using High Throughput Technologies

Samantha Burnham, Katarina Novakovic, Mark J. Willis, and Allen R. Wright. Chemical Engineering and Advanced Materials, Newcastle University, Merz Court, Newcastle-upon-Tyne, United Kingdom

The objective of High Throughput Technologies (HTT) is to allow a high number of experiments to be completed in the shortest time period possible, achieved by performing experiments in parallel using robotic workstations. HTT has shifted the bottleneck in process R&D away from screening to a problem with high throughput process development. One large pharmaceutical company recently reported(i) that 85% of the successful chemical syntheses emerging from automated discovery are currently rejected for commercial processes. The reasons for rejection are diverse, including economic viability, difficulties in downstream processing, process safety and environmental impact. As the number of compounds screened in the search for new catalysts, chemicals and pharmaceutical actives grows, the pressure to reduce this high rejection rate is increasing. To extend the application of HTT from discovery to product and process R&D is a multifaceted challenge involving many areas core to chemical engineering unit operation design.

In addition to the increase in laboratory throughput, by several orders of magnitude, advances have been made in software tools for molecular modelling in discovery and in the application of simulation technologies based on chemical engineering principles for design, scale up, optimisation and control. The integration of high-speed experimentation with novel experimental design procedures combined with statistical data mining, predictive modelling and simulation methods has the potential of delivering intensified process development (IPD).

Reaction mechanism is a prerequisite for the regression of kinetic parameters and the use of process modelling and simulation tools, providing a link between chemistry and chemical engineering. Although commercial software exists for kinetic fitting, the elucidation of reliable reaction mechanisms from complex networks of reactants and products remains a limiting step requiring considerable time and intellectual effort. This effort may be reduced facilitating IPD with the use of robotic workstations to provide data for automation of reaction mechanism determination.

Finding a useful model of a number of interacting chemical reactions (i.e. a reaction mechanism) is mathematically equivalent to a network inference problem, and may be redefined as the identification of the structure and parameters of a reaction network. Crampin et al. (2004) provide a survey of mathematical and computational techniques proposed to deduce complex biochemical reaction networks. Typically, these methods require the generation and analysis of significant quantities of experimental data in terms of composition-time series (e.g. Arkin and Ross, 1995; Arkin et al, 1997; Samoilov et al. 2001; Tsuchiya and Ross, 2001; Kremling et al., 2004; Bendtsen, et al. 2001). In particular, the methods based on system perturbations around stationary states do not lend themselves to the type of experimentation performed in most robotic workstations, i.e. lab scale batch reactors.

Crampin et al. (2004) suggest a general approach that appears suitable for batch experimentation. They suggest a methodology involving the selection of specific terms from global models comprising sets of basis functions, where each set corresponds directly to the set of all possible reaction rate terms involving the chemical species. Burnham et al. (2006) and Searson et al. (2006) propose and develop a similar approach that uses mathematical and statistical methods to iteratively select the basis functions and identify the corresponding parameters. However, this work did not take into account a number of practical constraints associated with robotic workstations.

The objective of this paper is to present an evaluation of the extent to which HTT experiments may be used to provide the necessary quantitative understanding of both physical and chemical phenomena required for IPD. This will be achieved using a robotic workstation (the Chemspeed SLT 106 synthesizer) and reaction calorimeter with the mechanism elucidation procedure proposed by Burnham et al. (2006) and the BatchCAD modelling and simulation software It is recognised that commercial automated jacketed reactor systems with vessel sizes of 50ml and above provide efficient stirring, accurate dosing and individually instrumented and computer controlled cooling. These are generally considered to provide reliable data for process scale up studies. Although small scale HTT equipment may have this capability, this does not necessarily translate into the capacity (rates of, mixing, mass and heat transfer, accurate dosing, sampling accuracy etc.) necessary to provide reliable data for process scale up. A case study using an L-proline catalysed aldol reaction will be presented to illustrate the work.

References

Arkin, A. and Ross, J. (1995) Statistical construction of chemical reaction mechanisms from measured time-series, J. Phys. Chem., 99, 970-979. Arkin, A., Shen, P. and Ross, J. (1997) A test case of correlation metric construction of a reaction pathway from measurements, Science, 277(5330), 1275-1279. Bendtsen, A.B., Glarborg, P., Damjohansen, K. (2001) Visualization methods in analysis of detailed chemical kinetics modelling, J. Comp. Chem., 25, 161-170 Burnham, S.C., Searson, D.P., Willis, M.J. and Wright, A.R. (2006) Towards the automated elucidation of chemical reaction mechanism from batch reactor experiments, AIChE Process Development Symposium, Palm Springs. Crampin, E.J., S. Schnell, S. and McSharry, P.E. (2004) Mathematical and computational techniques to deduce complex biochemical reaction mechanisms, Progress in Biophysics & Molecular Biology, 86, 77–112. Kremling, A., Fischer, S., Gadkar, K., Doyle, F., Sauter, T., Bullinger, E., Allgöwer, F. and Gilles, E.D. (2004) A benchmark for methods in reverse engineering and model discrimination: problem formulation and solutions, Genome Res., 14, 9, 1773-1785. Samoilov, M., Arkin, A., Ross, J. (2001) On the deduction of chemical reaction pathways from measurements of time series of concentrations, J. Chaos, 11, No. 1, 108-114. Searson, D.P., Burnham, S.C., Willis, M.J. and Wright, A.R. (2006) Identification of chemical reaction mechanism from batch process data, The 17th IASTED International Conference on modelling and simulation, Montreal, May. Tsuchiya, M., Ross, J. (2001) Application of genetic algorithm to chemical kinetics, J. Phys. Chem., 105, 4052-4058

(i)Private communication, Allen Wright, 2004

421d Enhanced (Multi-Throughput) Experimentation in Petrochemical Process Development and Troubleshooting

Joseph B. Powell, Chemical Process Technology, Shell Global Solutions US Inc., Westhollow Technology Center, P. O. Box 1380, Houston, TX 77251-1380

"Enhanced" or multi-throughput experimentation methods can greatly speed the screening of new chemistries and catalysts at the discovery phase of research and development, prior to extended testing in larger-scale units. The approach can also be applied after process development, piloting, and commercialization, to effect process improvements and optimization. A key feature of the latter is to have a tie point control for existing performance, as part of each test block. While high throughput methods (99 or more samples in parallel) may require extensive capital expenditure and time for implementation, it is often possible to engage in multi-throughput tests (2 – 12 or more) with similar investments and timing as required for one-at-a-time testing. Several generic examples of multi-throughput experimentation are provided for batch and continuous flow systems encompassing homogeneous and heterogeneous catalysis in batch and continuous flow reactors, liquid extraction, crystallization, flash distillation, and other unit operations.

421e A Novel Reactor for Rapid Scale-up and Commercialization of Heterogeneous Catalysts in Their Final Commercial Form

H. Sam Bergh, Stephen Cypes, Alfred Hagemeyer, **Nallakkan Arvindan**, Valery Sokolovskii, and Victor Wong. Symyx Technologies, 415 Oakmead Parkway, Sunnyvale, CA 94085

Since technology development in high throughput experimentation for heterogeneous catalysis discovery ramped-up in the 1990's, most of the reactor development has been in primary screening (100-300 catalysts in parallel for qualitative analysis) and secondary screening (8-64 catalysts in parallel for quantitative analysis). In addition to reactor technology at these throughput levels, efforts were made to develop corresponding synthesis and analytical tools to match the throughput of the reactors such that no additional bottleneck formed in the overall workflow. These primary and secondary screening workflows have matured in technology over the years for heterogeneous catalysis applications, and several companies, such as Symyx, now even offer these moderate throughput capabilities as products for-sale.

However, the rate of commercialization of discovered catalysts from high-throughput workflows has not increased at a rate comparable to the rate of discoveries resulting from secondary screening. This has continually been due to the bottleneck that forms at the point of pilot plant trials, in which the catalyst is formulated and analyzed on commercially-sized supports for the first time (secondary screening may be on commercial supports, but is typically screened as crushed and sized particles). Due to typically low availability of pilot plant equipment, and the expense and time associated with running a single catalyst in a pilot run, only several attempts may be made at the pilot scale before stopping commercialization efforts, even if secondary screening resulted in commercially-viable leads.

In this presentation, a novel reactor to screen heterogeneous catalysts in their final commercial form (i.e. beads, rings, cylinders, extrudates, etc.) on a benchtop in a rapid, high-throughput mode will be described. Efforts to design the benchtop reactor to match hydrodynamic and heat transfer characteristics of an industrial fixed-bed reactor will be discussed. As a case study, results for several different catalysts in a highly exothermic partial oxidation application will be shown, in which each catalyst could be fully characterized within 2 days at lab-scale conditions. This reactor, although not intended to replace a successful pilot plant campaign prior to commercialization, is a powerful tool to increase the likelihood of a successful pilot run on the first attempt. In addition, it can prove powerful for QC of catalyst batches from vendors or for process optimization studies.

421f Accelerated Physico-Chemical Characterization of Complex Protein Mixtures for Optimized Ion-Exchange Chromatographic Separation

Alex Berrill[1], Sa Ho[2], Daniel Bracewell[1], and Nigel Titchener-Hooker[1]. (1) Biochemical Engineering, University College London, Torrington Place, London, WC1E 7JE, United Kingdom, (2) Pfizer Global R&D, 700 Chesterfield Village Parkway West, Chesterfield, MO 63017

Development of chromatographic separation for complex protein mixtures such as crude extract or lysate is typically very labor intensive and time consuming. Chromatography sorbent selection and the buffers to use with them are traditionally made based on knowledge from similar processes and a large number of small-scale chromatography runs. This could require a significant quantity of feed material and extensive analytical support, which at early development stage are likely in limited supply or not available.

This study outlines the use of Surface Enhanced Laser Desorption Ionisation-Time of Flight-Mass Spectrometry (SELDI-TOF-MS) technology to obtain physico-chemical properties of proteins in complex mixtures, which can then be utilized to highlight promising chromatographic conditions for purification of a target protein. The key benefits of this approach over the traditional methods are the speed and ease of ProteinChip preparation and data acquisition to generate a large amount of quality data within a few days using only microliters of the complex protein solution. Data from the SELDI are processed to provide detailed knowledge of the bind/release properties of each protein in the solution under various buffer conditions. This knowledge will eventually enable to the selection of sorbents and buffer conditions that exploit the greatest difference in adsorption/desorption characteristics between the protein of interest and impurities in order to find the most effective chromatographic separation step.

421g Increased Efficiency in Chemical Process R&d by Automation - the Discovery of An Innovative and Extremely "Green" One-Step Synthesis of Tetronic Acid

Josef G. Schroer, Chemspeed Technologies, 7 Deer Park Drive, Suite L, Monmouth Junction, NJ 08852

Tetronic acid is an important key intermediate for different chemical products (e.g. (+)-Biotin and Losigamon) of the fine chemical and pharmaceutical industry. The industrial production of tetronic acid can be achieved in a multi-step gas phase process, starting from 4-chloro aceto-acetate. It was found, that addition of 4-chloro acetoacetate to a cold, aqueous solution of sodium hydroxide yields a mixture of sodium chloroacetate and sodium acetate. In addition, the resulting yield of sodium tetronate is less than 30%. Based on these findings, an innovative one-step process for the synthesis of tetronic acid was targeted. Therefore, the pH and temperature dependency of this reaction was studied in a parallel, automated set of experiments. This study resulted in an innovative and extremely "green" process, which saves several chemical steps over the pre-existing industrial synthesis of tetronic acid, and may yield to major cost savings in the production of this important intermediate.

This and other relevant examples, presented herein, impressively support the enormous value of a parallel and automated High Output Experimentation approach.

421h High Throughput Process and Materials Characterization Using Automated Td-NMR Technology

Supriyo Ghosh, minispec Division, Bruker Optics Inc., 2700 N Crescent Ridge Dr, The Woodlands, TX 77381

The manufacturing processes in chemical industries involve application of various unit operations on raw materials. These processing steps induce multiple physico-chemical changes leading to the desirable properties in the final product. During the process development as well as the industrial-scale implementation of the same, monitoring of the effects of each processing step need to be fast, accurate and precise. TD-NMR (time domain nuclear magnetic resonance) is a technique which can be utilized to look into the physicochemical processes at a molecular level and precisely quantify them, using a small bench-top instrument. It obtains the signal from the atomic nucleus in a material. Mathematical analysis of the TD-NMR signal reveals the changes occurring in molecular level in a material. Measurements can be done in a matter of a few seconds to obtain quantitative characterization of materials undergoing physico-chemical changes. The entire measurement procedure can be done in a continuous fashion using an automated robotic system. In this presentation, the basic principle of TD-NMR technology will be introduced. Various high throughput automated TD-NMR characterizations relevant to process development for chemical industry will be elaborated including but not limited to: extent of polymerization, crosslink density, crystallization, copolymer ratio, polymer density, finish on fiber, hydration, hydrogen content in fuels, impurity quantification, drying kinetics, fat-moisture-protein content, emulsion stability, droplet size distribution, porous media microstructure and transport. Probable new areas of chemical process industry where this TD-NMR technology can be exploited in the future will also be discussed.

CONTRACT MANUFACTURING – A VIEW FROM BOTH SIDES

By Kurt C Schneider

When the decision has been made to launch a new product, there isn't much thought given to where it will be produced. However, to get to the point of having a viable product **to** launch, many factors have to be considered. One of the most compelling factors to consider is whether to manufacture the product in-house or to have it contracted out. Depending on which side of that equation you are on, either the product developer or the contract manufacturer, the way you view this potential business partnership can lead to either a very smooth launch that is profitable for both parties, or can lead to strained relationships, sub-quality product, or a new product launch that is destined for the scrap heap.

In this article, we will take a look at that developer-contract manufacturer relationship from both viewpoints, and see if we can establish some common ground for you to take to your next project or negotiation. We also want to examine the pitfalls of a strained relationship, and give you some hints on how to recognize them, address them, and turn them in to positives for both sides. In effect, what we are dealing with is a negotiation, and your job is to ensure your 'side' gets a fair deal, but not at the expense of the other side. This is where a partnership with your developer or contract manufacturer is critical. As I have been on both sides of this equation on many different occasions, I believe I can provide a unique perspective that can benefit both sides. I can tell you right up front that honesty and the attitude you bring to the table will ultimately determine the success of your project.

To start, I'd like to take you through a real-life example of the process, as seen from both sides, the developer and the contract manufacturer. Once we get through the example, I'll give you the five most common reasons for failure in the process, and will then share some strategies for getting the most out of your contract manufacturing experience. I have changed the names and altered a few of the details to protect what others may feel is confidential information.

THE CASE OF THE STUBBORN DEVELOPER

Joe (not his real name) had been a product developer at a well-known and highly regarded Nutraceutical company for over 20 years. He has had many innovative product designs to his credit, and has commanded a wealth of respect not only internally but also within the Nutraceutical industry. In fact, his reputation in the industry preceded him. Unfortunately, along with this high level of accomplishment came a level of stubbornness when faced with a new idea. So, when Joe was given the task of developing a new product with no previous testing, a product that had never been on the market, it was a challenge he was ready for. With market data to support the potential for a product such as this, Joe's boss saw it as an opportunity to make a mark on the industry, and firmly establish his own name as a force in the Nutraceutical industry. It was a natural fit for the star developer to take on this project.

That was over a year ago.

Today, Joe has a good idea of what the product should look like and act like, and has a good bench scale working prototype. He also knows the current production in his facility was at capacity. Knowing the huge potential, Joe decided to place a few casual phone calls based on some recommendations of colleagues, and sets up an appointment with me, a contract manufacturer specializing in, you guessed it, novel Nutraceutical products.

Joe's first meeting with us involved our Director of Contract Manufacturing, with the company President there as a gesture of good will. What happens next should have sent off warning signs to everyone in attendance. After the formalities of introductions and handshakes, a 'Blanket Confidentiality Agreement' was signed by our President and Joe on behalf of his company. Now that our collective hands were tied, Joe began his presentation on the history of the new product, explaining features, benefits, design, materials used, and, finally, in the final ten minutes of his 45 minute presentation, a brief summary of his prototype and bench top production process. Of course, with the 'can do' attitude that is the motto of our company, the Director of Contract Manufacturing states unequivocally, 'no problem Joe, we do this kind of stuff every day!'.

Thus starts the downward spiral of miscommunication, over-promise-under-deliver, cost over runs, missed deadlines, and an end result product that not only does not match the bench top prototype, but also has a specification wide enough to drive a truck through.

Sound familiar? If not, let's look at it from the other side:

THE CASE OF THE CAN-DO CONTRACT MANUFACTURER

Dave (not his real name) was the Director of Contract Manufacturing at our company, and he is under tremendous pressure to close some big sales before their year-end. It's been a lean year, and the phone call from Joe was just the kind of energy boost Dave needed. With all the excess capacity, our company was staring down the barrel of a multi-million dollar loss for the year, and Dave was the man on the bubble to fix it. He knew he needed to generate some revenue, but more important, he needed to secure business for the following year. In short, Dave was desperate.

Dave had been in the contract manufacturing business long enough to know a few tricks to gain business, and was fully prepared to use them in their meeting with Joe. No matter what Joe showed them or asked them to do, the answer was going to be 'yes'. Even if they had no practical way to manufacture the product, which proved to be true, Dave was going to say 'no problem'. Besides, he had done most of his biggest deals this way, and had really put it on R&D and Manufacturing to work it all out, something they did regularly, but not without collateral damage.

Dave took a moment to stop by the President's office to invite him to the meeting as a gesture of good will. He told him who the customer was, how big the potential volume could be, and the bottom line dollars it could add to the company if it was costed 'correctly'. Seeing the look on Dave's face concerned the President. He had seen that look before, right before the company had a massive recall due to a microbiological contamination. The outbreak, it seemed, came from running the process too fast, introducing a level of uncertainty that had not yet been tested. That little 'risk' cost the company over 2MM, and cost the previous Director of Contract Manufacturing his job. Dave was hired to bring a better sense of ethics and honesty to that group. "He couldn't be feeling that much pressure that he would consider agreeing to manufacture something he knows we couldn't do? "thought the President. He let Dave continue on to the meeting.

In the meeting, after Dave got the CDA signed, he sat back and listened, sort of, as Joe gave his presentation. Since Dave was already set with his 'no problem' answer, he really didn't need to hear any of this, did he? Besides, he had heard it all before, and he had even heard about Joe, that he did his homework and had a track record of some really inventive products. If Joe was that good, and he knew his R&D and Manufacturing personnel were always up for a new challenge, then this was the perfect opportunity.

The perfect opportunity to over-promise-under-deliver, miss deadlines, have cost overruns, and an end product that does not meet the original specifications.

If you haven't already figured it out, I have played the part of both Joe and Dave. I have been the stubborn product developer, knowing everything about my product but communicating little, and I have been Dave, under pressure to deliver sales, and backing our company in to a corner promising a product we could not deliver.

What went wrong?

FIVE WAYS TO ENSURE FAILURE

If Joe and Dave did not invent the following list, then they must have had a great deal of input in creating it. Here, then, are the five major ways to ensure failure in the contract manufacturing process:

1. Throw 'due diligence' out before you start

Joe did not research contract manufacturing options, just taking the first one who answered the phone. While you could get lucky with a contract manufacturer who has a good reputation and will conduct it's own due diligence, chances are this is the first step down that slippery slope. There are many companies out there that have tablet, capsule and softgel production capabilities, and it is up to you as the developer to find them all, large or small, and compile a list of those companies that have the equipment to do your product. Not necessarily the process you feel is the right one, mind you, just make sure they have the equipment they say they have. The Thomas Register is a great resource for these types of searches.

2. Leave engineering to the train conductor

Joe is a developer. The task of working with a contract manufacturer is best suited for an R&D engineer. They speak the language better. At the very least Joe should have invited someone from the process development team and/or their own manufacturing department to get in early on in the discussions. I have worked with companies where the product and process developer form a two-person team for every project that is ready for scale-up and plant trials. This is an excellent way to merge the two areas of expertise. One of the best examples of this in action was a project I had worked on as the Process Development Engineer. We had a three-person team with a Product Developer and a Purchasing Agent. The three of us were empowered to make all decisions from the product aspect to make the launch happen on time. I worked directly with the contract manufacturer, and served as a liaison between the product developer and the manufacturer. To this day I feel this is by far the best way to develop and launch a product, especially one with new technology.

3. Always tell the developer you can make the product better/faster/cheaper

Our contract personnel argued with Joe as to the best way to manufacture the Nutraceutical product. Since we had been in the business for years, when we met Joe, it was only a matter of minutes before we began to explain that we would be able to put together a better process and make a better product than Joe specified. Of course, this put off Joe almost immediately. Listening, not arguing, is the only way to make this relationship work. Joe had worked long and hard to develop a product that had all the properties the Marketing Group wanted. To hear us argue that we knew how to make it better after 30 minutes of discussion nearly killed the deal before we even had a chance to test it. Both parties should be open to new ideas, agreeing on what is critical to maintain and what is flexible in both product and process is what matters.

4. Make sure you rewrite the customer specification to best suit your manufacturing needs

Unfortunately, some contract manufacturers develop their own internal specifications for the product, as they feel it will make a 'better' product. While they met all of Joe's specifications, which were quite wide to begin with, they also added things which created an entirely different product. This one never ceases to amaze me. No matter how specific you are in your requirements, some contract manufacturers will attempt to make as many adjustments and shortcuts to your product, all to save a penny or two and improve their bottom line. There are some times when internal specifications are positive, such as when they are tighter than the customer requirements. In that respect you are maintaining their target product and are showing good will by trying to make the best product within their specifications.

5. Always do the first run to help accounting cost out the product

By doing this, and making a great product, a contract manufacturer can pretty much ensure they will keep your business, no matter what price the eventually quote. If a contract manufacturer cannot give you a very accurate quote based on some bench and/or pilot trials, at no or minimal cost to the customer, you may end up being on the wrong end of this trick. If a contract manufacturer says it needs to run a large scale plant trial to cost your product accurately, and they want you to pay for the run and materials, you are best to look for another manufacturer. I have seen this, especially in cases where there are only a few manufacturers who can make your product. The less choices you have, the more leverage they have.

HOW TO GET THE MOST OUT OF YOUR CONTRACT MANUFACTURING EXPERIENCE

1. Conduct Due Diligence

Both sides of the Developer/Contract Manufacturer equation must have thorough understanding of what it is they want and who can provide it. If you are the developer, never be satisfied with a recommendation of a colleague as your final selection decision. There are plenty of ways to find any and all contract manufacturers who can mass manufacture your product today, as well as those who say they can, but not without some substantial capital changes. If you are the contract manufacturer, research your potential customer. Are you bidding to be a second-source manufacturer, or is this a new product line? Is the developer working with any other potential manufacturers? What about the developer's company? How big are they? Do they pay their bills in a timely fashion? Are they accessible for information after the initial start-up operations? The more you know about each other going in, the better prepared you will be to get what you need.

2. Communicate early and often

Yes, communication is one of the most overused words in the corporate dictionary. However, in the case of the developer/contract manufacturer relationship, having successful meetings and bench tests early on are the only ways to get started on the right foot. You should have a kick-off meeting with at least three members from the developer's team and the contract manufacturing team. A suggested line-up would include;

- Customer
  - Developer
  - Marketer
  - R&D Process Engineer
  - Developer's Production Plant Management
- Contract Manufacturer
  - Contract manufacturing department head
  - R&D Process Engineer
  - Manufacturing Management

This list is not all-inclusive, but should be the bare minimum for a successful kick-off

3. Go and watch a batch of the product being produced at the customer's facility

This is a most critical step. Even if it is only a bench sample batch, the value of seeing what it is you will be making, and, more importantly, what it is they expect to see, is knowledge and information that cannot be transferred over a conference room table. If the product is unique and has not been mass-produced, then I would suggest multiple meetings at the customer site to view any bench activities, and get to understand the thinking of the developer. If you understand the mindset, you are well on your way to helping them achieve their goals, and create profit for your company.

4. Agree on all aspects of the specifications before any batch is run

As a developer, be prepared to discuss any and all your specifications for relevance. If a specification is not relevant, now is the time to drop it. You will show good will towards the contract manufacturer by showing them you are willing to compromise.

One of the most trying points of the negotiation, I have found, is defending what has turned out to be a subjective specification, such as 'to match standard' or 'typical of....'. If you truly have a standard, then make sure you have some with you at the kick-off meeting, and that you can reproduce that standard whenever the contract manufacturer requires it. There have been many occurrences where subjective specifications such as color have drifted over time that what is considered acceptable today would have been rejected using the original standard. Don't fall in to the spec drift trap. Update your standards often, such as yearly, and ensure all your contract manufacturers have them.

5. Get a firm quote on the business upfront

It means toll charge, development costs at the contract manufacturer's site, and any hidden costs, like freight, rush orders, or extra testing. It means not only a quotation from the contract manufacturer, but a volume commitment from the developer. This is why having that Marketer at that kick-off meeting is so crucial. Seldom does a Developer have a true understanding of the market potential of a new product, and seldom will a Contract Manufacturer quote a piece of business without a firm volume number. Sounds simple, but is one of the hardest negotiation points. How many times have you, as a developer, been asked by a contract manufacturer, "What is the volume potential?" and, stuck from lack of knowledge, give some unreasonable high figure only to watch the manufacturer's eyes bulge? How about the other side? How many times has the developer made the statement, "can you give me the rough cost to do.........." knowing full well your answer will either make or break the possibility of getting this business.

You are both stuck, so why not do the homework first, and come to the table knowing these figures. It would save both parties numerous headaches and backtrackings.

6. Be there for the first production run(s)

There is no way a contract manufacturer, even if you have done all the work above, will be able to make your product right the first time, 100% of the time, right at target spec. You need to be there to help QC and Manufacturing personnel interpret what they are seeing on the line, help them make adjustments, and help them understand why the specification you have set that they are struggling with is so critical to the product's success. You should be there to collect samples, data, and be able to communicate at all levels of the contract manufacturers operations so everyone understands what is expected.

SO THAT'S ALL THERE IS TOO IT?

Sounds easy, right? Well, the toughest part is still to follow. After a contract manufacturer is up and running your product, as a developer it is not enough for you to walk away and move on to your next challenge. That's where spec drift will bite you. A contract manufacturer, through no fault of their own, will allow production to slide towards the end of a particular specification that gives them the greatest throughput with the least amount of trouble. You may see this one of two ways; either the manufacturer is being lazy, or a sign for you to review your specifications and ensure the product at that end is acceptable. Remember, you negotiated for that specification, and any product produced within that specification, per the understanding of the contract manufacturer, is acceptable. If you call this product in to question, you risk undermining the entire project. Don't get upset, review your spec.

You will also need a solid follow-up plan, with sample reviews, on-site sales updates and volume projections. These meetings will ensure the continued success of your contract manufacturer, and will give you a relationship that will go well beyond the boundaries of that product launch. Building trust and confidence in each other will help in future projects, cutting down the initial team-building discussions and saving time in development and launch.

IN SUMMARY

The Developer/Customer – Contract Manufacturer relationship, in its purest form, is a negotiation. If both parties come in to the initial kick-off meeting prepared and understand what they can and cannot do, the process is off and running and you will both be rewarded with a successful product launch. Upfront communication is the key, but staying with the process, with both parties having daily involvement through start-up, is critical. Following-up with the manufacturing, doing product reviews and adjusting product specifications where appropriate, will not only guarantee the ongoing success of the product, but will also cement a relationship and widen your network.

477b Bringing New Chemicals To Market - Outsourced Process Development

Jeffery Perl and Stanley Yoslov. Chicago Chem Consultants, 14 North Peoria, Suite 2C, Chicago, IL 60607

The decision to outsource your chemical manufacturing has been made. Now your hard won process must be transfered from your lab to the Toller where it will be developed for manufacturing. Along with this are challenges of quality control and economics as well as timing of raw materials and delivery of product. This paper will look at the items of consideration necessary to success.

477c Contract Manufacturing In A Global Environment

Mike J. Doll, HODer, Shell Global Solutions, 3333 Highway 6 South, Houston, TX 77082 and Jonathan H. Worstell, HODER, Shell Global Solutions US, 3333 Highway 6 South, Room E-2061, Houston, TX 77082.

Contract manufacturing can be a challenging endeavor. Over the span of years the rules and challenges have become "routine" as we learned the ropes. With the advent of Globalization, we have expanded our searches for new contract manufactures to overseas locations. Are the rules and challenges still the same in global environment? This presentation will touch on items we knew and were comfortable with, but more importantly, with our recent experiences in global contract manufacturing.

477d Bringing New Chemicals To Market - Regulatory Considerations

Jeffery Perl, Chicago Chem Consultants, 14 North Peoria, Suite 2C, Chicago, IL 60607

ABSTRACT Bringing New Chemicals to Market – Regulatory Considerations

USEPA regulations require submittal of a Pre-Manufacturing Notice (PMN). The PMN process allows EPA to understand the nature and toxicity of the new material and how it will be manufactured and regulated. This paper will examine EPA and OSHA considerations as well as a look at Global implications.

477e Technology Innovation: Transitioning from Hunter-Gatherer to Cultivator

L. Burr Zimmerman, Brian E. Henslee, and Robin Ng. Techriculture, 5797 Stallion Drive, New Albany, OH 43054

Turning technologies and inventions into commercial successes is no easy task. Most commonly, IP holders seek to match existing technologies with existing licensees while minimizing investment and risk. Refining innovations to match market needs or finding secondary applications may often be overlooked or avoided. To realize the full societal and commercial potential of research, we must not rely solely on gathering promising technologies as they emerge. Instead we must cultivate innovation by engaging and fostering ideas from their earliest stages.

Techriculture is a non-profit organization focused on cultivating technology innovation. We form project teams comprised of PhD and MBA students to work directly with researchers and in close coordination with technology managers. Teams are formed with a unique combination of skills: deep technical subject understanding and entrepreneurial and commercialization experience. The teams work directly with researchers not only to develop primary market applications, but also to seek novel new ways to utilize the researchers' underlying technology and skills in the marketplace. Techriculture helps researchers and technology managers establish a significantly more valuable position from which to develop commercialization strategies.

Educational outreach is a critical component of Techriculture as well. Numerous outlets have reported the need for not only training in math and science, but also in technical innovation. Targeting students of all ages and working closely with entrepreneurship and educational organizations, we provide training and outreach programs to enhance students' scientific and entrepreneurial mindset. Techriculture is working to increase the number of young people interested in being a part of technology startups and other technology commercialization efforts.

Techriculture

Cultivating technology value

Web Page: www.techriculture.org

525a Uop Pilot Plants – Developing Technologies And Careers

Guy Woodle, Kerry Crosby, and Daniel Pintar. UOP LLC, 25 East Algonquin Road, Des Plaines, IL 60017

UOP LLC, a Honeywell company, has been developing technology for the petroleum refining, gas processing, and petrochemical industries for over 90 years. One key to UOP's success is a fully functional Experimental Development Department (EDD) that is accountable for design, construction, and operation of pilot plants. UOP's EDD is closely aligned with the Development Department, which takes information generated in pilot plants and transforms it into viable commercial plant designs. Numerous pilot plant and process development career opportunities exist at UOP.

UOP's EDD designs, constructs, and operates pilot plants for product development, process development, and commercial support. Specialized skills are required to meet this diverse pilot plant demand. Pilot plant engineers must understand the technique of scaling-down process designs to ensure information collected on the pilot scale can be applied to the commercial scale. Other key skills for pilot plant engineers include reviewing experimental design, developing operational improvements, and advancing test methodology. At UOP the majority of pilot plant operation is carried out by non-degreed employees who have developed specific, unique skills and knowledge in the equipment and operation of pilot plants. These unique skills are developed on the job and are critical to the success of the organization.

525b A Career Working For The Process Development Company

Joe Schroer, ClearWaterBay Technology, Inc., 4000 W. Valley Blvd., Suite 100, Pomona, CA 91789

Process development is a multi-functional, multi-national, multi-objective career. This presentation describes the career working for a small private process development and consulting firm. Traditional chemical engineering training and skills are needed to perform tasks for the process design, modeling, and experimentation cycles during process development, for example the evaluation of fuel chemical processes or utilization of alternative feedstocks. However new skills must be constantly acquired, for example understanding new chemistries or chemical production from biological feed sources. Creativity and ingenuity are also used to improve the intellectual property positions of clients in their competitive areas as well as to implement the business and marketing strategy of our own company. Usually projects are multi-organizational collaborations, where our members simultaneously meet with chemical companies, engineering & construction companies, other consultants, industry consortiums, universities, etc. for the development of a chemical process. Opportunities for career satisfaction come from a variety of aspects, intellectually, both from technical and business challenges, interactions with a variety of people, and development of your own creative inventions.

Web Page: www.cwbtech.com

525c The Value Of A Pilot Plant Experiences After Graduate School

C. Michael Smith, Freeport, TX, Dow Chemical, 2301 Brazosport Blvd, B-4810, Freeport, TX 77566

Upon completing graduate school, young researchers often avoid experiences involving pilot plants because of preconceived biases. Pilot plant research may sometimes be perceived to be mundane, routine, or lack fundamental science. It is, however, neither mundane nor routine, but often provides opportunities to develop and exhibit personal and technical leadership at an early stage in a researcher's career. Many problems encountered in pilot plant research require either the development or the application of process fundamentals. It is through this solution process using fundamental science that a number of skill sets are developed leading to a rewarding career. Examples of these career paths will be discussed.

525d The On-Going Reinvention Of Process Development

Roger L. Parsons, Process Development, The Lubrizol Corporation, mail stop 112A, 29400 Lakeland Blvd, Wickliffe, OH 44092

The reason for Process Development remains "to build technical advantage into a company's overall strategy." But how we do that is undergoing profound changes under such influences as globalization, intensified competition, shifting demographics, evolving corporate cultures, and the on-going development of the information age.

Corporations face mass brain trust retirements among baby-boomers, declining (non-IT) engineering and science graduates, and stock market pressure to grow the business with the same, if not fewer, employees. Nowhere is new technology optional; "reinvent (or at least improve) or die" is a simple fact of life for any technology-based corporation.

To rise to these challenges, we have: shed past hierarchical cultures, adopting instead self-directed work teams; experimented with faster / more flexible ways of sharing our process knowledge gains; worked to better understand and facilitate innovation (e.g., developing differing pathways for new product development suited to differing risk profiles); and honed our basic skills.

This talk will review these trends from the perspective of The Lubrizol Corporation's transportation additives business, citing both gains made and dead-ends left behind. In particular, this talk attempts to answer: "How does Process Development maximize its contribution (to the corporation)?" Special focus will be given to: the role of the process development engineer; the training foundation needed to enable process development engineers; rising to the unique challenges of safety in pilot plants; and how we manage our chief product – knowledge – as the speed of information exchange ever increases.

Hardly a finished story, this talk will close with a summary of the challenges and promises ahead.

The basic themes of the talk are: "balance" over "magic bullet" solutions; the on-going evolution, if not reinvention, of how a Process Development Department makes its impact; and an attempt to frame this flux we are in, and the possible paths ahead.

525e A Career in Process Development at Shell

Joseph B. Powell, Chemical Process Technology, Shell Global Solutions US Inc., Westhollow Technology Center, P. O. Box 1380, Houston, TX 77251-1380

A career in process development provides opportunities spanning idea generation and scouting research, all the way through commercial startup and subsequent plant optimization and support. Challenges encompass the entire array of chemical engineering disciplines and unit operations including reaction engineering, separations, process control, process synthesis and modeling, process economics, as well as the sciences of chemistry and catalysis, materials, fluid flow, thermodynamics, heat and mass transfer. Both experimental and modeling skills are valuable, as are interpersonal and business skills needed to lead a project team and influence senior management in decision making. Personal and technical skills can be expanded throughout one's career, by filling in the unresourced "white spaces" that inevitably arise during project execution. Given the need for extensive contacts across the organization to commercialize new technology, a process development career can mature into a position of influence across the organization.

572a Leveraging Pilot Plants Through The Use Of Models

Salvador Aldrett, Reactive Hazards Assessment, Shell Global Solutions US, 3333 Highway 6 South, Room E-2061, Houston, TX 77082

Throughout my experience in the modeling world, I have found that models without pilot plants cannot exist, and pilot plants without models take a longer time and effort to be optimized. Models provide an invaluable diagnostic and optimizing tool for pilot plants. On the other hand models need data, and you model will be only as good as the data you input, or as the old saying goes: "garbage in, garbage out." This paper presents a series of examples that highlight the synergies between models and pilot plants. The common element is that aspen was used as a simulating tool, but that is where the similarities stop. Modeling can be used in different pilot plants, for different process, and unit operations. The common thread through all these cases is that in every single one, pilot plants, and ultimately processes were optimized through the use of simulation.

BEST PRACTICES IN PROCESS DEVELOPMENT OF NEW TECHNOLOGY
BEST PRACTICES IN PROCESS DEVELOPMENT AND PILOT PLANTS
AIChE 2007 FALL PROCEEDINGS

Mary Ellen Yarossi, IPA Institute, Ashburn, Virgina

As businesses seek growth and profitability improvements, they often look to new technology to provide new business opportunities for growth or to improve their competitive position. While offering high rewards, these ventures can carry project risks from technical uncertainties as well as market risks that come along with the introduction of new products. In this paper we discuss a new Technology Development Planning Model that highlights practices shown through quantitative analysis to mitigate the risks arising in projects incorporating new technology.

The Database and Research Methodology

The research presented is based on Independent Project Analysis' (IPA's) detailed databases of capital projects, which contain about 10,000 projects in total. Each project is described in detail with a minimum of 2,000, but as many as 5,000, individual data elements, depending on the project characteristics and complexity. The databases contain detailed project histories collected from projects of all types and sizes executed all over the world. All major geographic regions of the world are represented, which allows for comparisons of regional differences. The size of the projects ranges from US$50,000 to over US$5 billion. The databases are segmented by project type, allowing for unique aspects of various industries to be quantified and related to performance.

Data are collected in all key areas to understand the project, including project characteristics; cost, schedule, and operational histories; safety practices and results; management strategies; project definition; resourcing and team integration; and project execution practices.

These data are used to develop a set of statistical models that can benchmark a project's cost, schedule, and functionality. This allows a measure of the competitiveness of a project relative to projects with similar characteristics. The impact of key project practices on outcomes can then be measured. For this study, we relate the practices associated with the definition, development, and execution of commercializing new technology to the results of these capital investments.

Innovative Projects Database

Detailed histories of research and development (R&D) efforts are collected for 850 innovative projects. These projects range in size from US$500,000 to US$2 billion. Various industry sectors are represented, including minerals, refining, commodity and specialty chemicals, pharmaceutical, and consumer products. Many of these projects construct new facilities, but the database also contains innovative projects that add to or modify processes at existing operations. The projects are located all around the world, although 63 percent were executed in North America and another 21 percent in Europe. The database also contains innovative projects from Asia, Australia, the Middle East, Africa, and Latin America.

Data on scale-up projects and piloting activities are collected. The type and amount of research and design are recorded. Characteristics of the development program as well as the types of prototype and testing facilities are included in the database. The level of difficulties encountered and/or expected early in a project in 18 separate process areas is rated. Problems in these 18 areas that were actually encountered during startup and early operations are also rated for severity. Company and industry experience is considered in assessing the level of innovation of a project.

Quantifying the level of innovation is important in understanding the appropriate technology development strategy and practices needed for success. For each project there are several measures of technical difficulty. These include the following:

- **New steps**: major functional blocks in a process that incorporate technology that is new to the industry. By "new" we mean new chemistry, new match of chemistry and equipment, or both.

- **Percent new**: the percentage of the total capital cost of a project (without capitalized costs of startup) associated with the new technology portions of the facility.

- **New integrations**: a first time, in the industry, integration of process steps that are otherwise practiced commercially.

- **Complexity**: the number of continuously linked block steps in a process required to perform all chemical and physical operations for the production of product. An average project in the industry has five to six block steps.

The database of innovative projects includes projects that range from a low level of innovation to a very high level of innovation. This distinction is important in understanding the technology development strategies and risks associated with the commercialization efforts. Developments for both new and existing products are included in the database. The scale used is as follows:

Revolutionary Innovation	**Pioneer process** – entirely new route to an existing chemical or new process for the production of a new product **Substantially new process** – analogous process maybe in use, but this is a new route or approach—core technology is new and the effects of the new core technology cannot be isolated **Major process modification** – two or more steps on the main process train are new in commercial use, but the core technology is unchanged or changes can be isolated from upstream and downstream effects
Evolutionary Innovation	**Minor process modification** – one or more steps in the process are new, but the order and function of steps remain unchanged from current practice

Off-the-Shelf Technology	**First-time integrations** of proven processes **Near-duplicate** of existing process

The range of projects and the depth of data for each project allow for quantifying the relationship between technology development practices and project outcomes.

New Technology Projects Perform Differently

The track record for new technology projects is dismal. In evaluating detailed histories of innovative projects, we find that 40 percent of the projects commercializing some level of new technology fail; that is, they never operated as intended and were considered business failures. Less than 20 percent of new technology projects delivered what was promised at authorization. On average, new technology projects experience cost growth of 30 percent and schedule slip of 65 percent. This means that a project estimated to cost US$25 million at authorization would have cost US$32.5 million and would have started up almost a year later than planned. Startup, the time from feeding of materials to steady-state operation, is also much longer for new technology projects. Each new step can add several months to the startup duration. Of particular concern is that new technology projects failed to operate as expected. Operability, which is a measure of the amount of product made in the first year of operation, can be significantly reduced when new processes are commercialized. This lack of production can reduce the return on investment (ROI). Processes new to Industry typically produce about 80 percent of their nameplate capacity. Note that we measure operability in the second 6 months of operation to allow for minor fixes and startup problems to be resolved. Some of the issues we see affecting operational performance include malfunctioning equipment, chemical reactions that do not produce expected yields, and unexpected process impurities.

These results lead to disappointment in new technology projects and have discouraged organizations from commercializing innovations. However, there is a group of projects that are very innovative and perform very well. These projects experienced cost growth and schedule slip comparable with that of standard technology projects. In addition, and most importantly, these projects do not have operational difficulties so the needed quantity and quality of product is available to meet business needs. The practices used by these excellent performing innovative projects are discussed in detail in the Technology Development Planning Model section below.

Technology Development Planning Model

The new technology planning model is based on four practices that have been found to be common across the best performing innovative projects. The practices are as follows:

1. Recognize the Degree of Innovation
2. Develop Basic Data
3. Resource the Effort Adequately
4. Ensure the Project is Well Defined

We discuss each of these practices in detail.

1. Recognize the Degree of Innovation

This may seem obvious at first, but this is one of the most common mistakes made in new technology projects: the fact that the project does incorporate new technology and therefore will be more difficult to execute needs to be recognized. As the results for cost growth, schedule slip, and operability showed, most projects set targets that are more in line with those of standard technology projects and then miss those targets.

One of the practices we find necessary for the successful implementation of new technology projects is to correctly identify the degree of innovation. Not all new technology projects carry with them the same amount of risk. For each new step, operability is reduced by 14 percent. For facilities incorporating more than three new steps, the change of failure is much higher.

Correctly assessing the level of difficulty is important to setting reasonable targets for cost and schedule and, most especially, operability. It also has implications for the type of development program that is needed to demonstrate the technology. Processes that are deemed more difficult require an integrated pilot, higher contingency, high-quality Front-End Loading (FEL), and team continuity through startup. Processes that are considered technically simply still require team continuity and high-quality FEL, but may not need an integrated pilot plant. A process is technically difficult if it:

- Is continuous
- Has four or more process steps
- Incorporates recycles
- Has a solid feed
- Introduces new technology in two or more of its process steps

2. Develop Basic Data

Every project is engineered from a Basic Technical Data package that governs the design. Examples of basic data included reaction yields; feedstock characterization; separation efficiency; and production of impurities, energy, and mass balances. For standard technology plants, the basic data are available for commercially operating process units. By definition, commercial units do not exist for an innovative process. Basic data then have to be obtained from analogous units, process development units, or integrated pilot facilities. The completeness of the basic data packages has been shown to be one of the most critical aspects of the successful commercialization of an innovative process.

Analyses of innovative projects show that batch processes carry much less risk than continuous processes. Therefore, for batch processes, the new technology can be isolated and tested as a single step in the process.

More complex processes that commercialize new technology require more extensive development efforts, and often an integrated pilot facility is necessary to obtain the needed process development information. By integrated pilot we mean a facility that has all major process steps as well as recycle streams. These more complex projects tend to have impurity issues or difficulty with recycles. The larger number of steps means that any difficulty may impact more process steps downstream from the new technology steps, and more complex processes are more likely to have more than one new process step.

An integrated pilot run for an extended period of time can dramatically improve the early operability of new technology processes. Revolutionary new technology projects for which an integrated pilot facility provided basic data operability in months 7 to 12 after startup averaged 79 percent of design capacity versus only 30 percent for a comparable process for which a pilot facility was not built. With a pilot facility, operating conditions can be fully explored and optimal operating ranges established. In many cases, the pilot plant can be used to train operators and learn about operations so that operating instructions can be established. Moreover, it is important that the pilot use feedstocks that will be used in commercial operations. In this way, the impact on production rates can be fully understood. As with operability, experience gained in running the integrated pilot facility reduces startup time significantly. The projects that had an integrated pilot plant achieved industry average startup times, paying no penalty for the incorporation of new steps. This is primarily because operating problems were identified in the pilot, and solutions to those problems could be identified prior to startup of the commercial facility. The capital investments for which a pilot facility was not built had startup times that were twice as long as industry average, sometimes taking over a year to startup.

Research has shown that the scale factor of the pilot to the commercial unit is less important than the fact that the pilot truly represents the commercial facility. The learnings from integrated pilots are critical for truly understanding how the process behaves. Several other important elements of the running of the pilots were also found to be common in the successful innovative projects in the PES Database, which are listed below:

- They were run in the same mode of operation as the plant, meaning they were run continually if the plant was to be run continually. This allowed for extended runs in which product was made. The effect of recycles, impurity build-up, catalyst poisoning, and side reactions could also be seen.
- Processes were tested thoroughly to explore chemistry and process issues. When process testing was abbreviated simply to demonstrate that the reaction would work, the process was not well understood and the commercial facility experienced startup and operability problems.
- Piloting and testing was best done at a central facility. This has been shown to allow for more thorough testing.

3. Resource the Effort Adequately

The first practice we discussed was to recognize the degree of innovation and the risks associated with implementing a new process. Part of this recognition is that new technology projects need to be resourced differently, assuring adequate representation from R&D and a technical specialist. Integrated teams, or a team with representation in the definition stage from all key functions, are a critical component to success for all types of projects. Innovative projects, however, may require additional resources from R&D and additional technical specialists. In the most successful new technology projects, all needed functions were represented. For some large and complex projects, team members were removed from all other ongoing work so they could focus on the innovative projects. Team building exercises were also common in the successful innovative projects.

Clear alignment of goals is also critical to new technology projects. This includes a good understanding of the needs of each function. In some of the successful innovative projects, teams mentioned that they felt that they were "on a mission." Extensive communication through frequent team meetings was the norm in successful innovative projects.

Continuity of the project team is another important element of success. Turnover of key personnel is a serious problem when commercializing new technology because key data may be lost with the change in personnel.

As the demographics in the process industry shift and a lot of experience is lost due to retirements, it becomes harder to resource projects adequately. This is especially true for new technology projects. One of the problems caused by the lack of experience is that the difficulty of new technology projects is not recognized. As a result, projects are started without the basic data developed and appropriate testing completed.

Recognizing the time needed for development efforts is import. More time is needed than for standard technology projects. The development of basic data and for definition cannot be rushed. New technology projects take longer to define and execute. For moderate levels of innovation, the time for project definition may be only 2 or 3 months longer than for standard technology projects. For technically difficult projects, the time for definition increases by almost 6 months, on average, relative to the same size standard technology investment. Engineering costs are also found to be higher by several percent as more engineering resources are needed for new technology developments. Resource needs are also higher in startup when technical specialists may be needed.

4. Ensure the Project is Well Defined

All projects benefit from being well defined at the time they are authorized. However, defining the project well is especially important for innovative projects. These projects often encounter more changes due to unknowns in the technology. Elements of project definition of special importance to new technology projects include the following:

- Thorough definition of the process including:
  - P&IDs
  - Heat and material balances (H&MBs)
  - Definition of feedstock characteristics
  - Process control strategy
- Early establishment of startup plans, including:
  - Staffing requirements to include technical specialists and vendor support
  - Turnover sequencing
  - Trial run strategies

- Preparation of integrated execution plans
  - Networked, resource-loaded schedules
  - Critical path defined
  - All major tasks defined

Although thorough definition is important for all projects, new technology projects are less forgiving, which means incomplete definition affects innovative projects more. Higher levels of innovation result in less predictable cost performance; thus, more contingency at authorization is needed to prevent an overrun. A standard technology project with a *Best Practical* level of definition at the time the project was authorized would require 5 percent contingency to prevent a cost overrun, whereas a project with 25 percent new technology would require 12 percent contingency. As the quality of definition degrades, the need for contingency increases so that same new technology project with an industry average or *Fair* level of definition at the time it is authorized requires 25 percent contingency. When the basic data are not complete, the cost growth is much higher. A contingency of 40 percent is needed for the same new technology project with a *Fair* level of definition, but incomplete basic data, because the lack of definition and basic data result in changes to the equipment design later in the project, which drive up costs. An important finding here is that some risk can be mitigated with good definition.

As we saw with cost predictability, the same relation holds with schedule predictability. With innovative projects, schedule slip often results from design changes as the technology development continues. As engineering deliverables slip, construction must be delayed. In some cases where the R&D may overlap project execution, design issues are not resolved until the project is in the field, resulting in longer construction times.

Some Observations on New Technology Projects

In the last 15 years, very few of the project systems in IPA's database have done any new technology projects. This means that many of the more experienced project professionals and business managers have never seen a new technology project. Recognizing and addressing the risk is critical. Innovative technology is a "burden" that project organizations are typically neither prepared nor able to deal with adequately. Often the proportion of new technology is underestimated, which, in turn, means the risks are underestimated. These projects often result in business disappointments.

While new technology projects do carry with them additional risks, our data have shown that these projects can be done effectively. Businesses attempting to execute innovative projects need to implement a set of Best Practices that can help achieve project success. When new technology is successfully commercialized, facility costs can be reduced, new products can be introduced, and operational improvements can be realized. To succeed in commercializing new technology projects, there must be a willingness to understand these projects are different, staff them accordingly, define them thoroughly, and do the necessary process development to provide complete basic data.

572c Operation Excellence in the Pilot - "Key Elements for Success"

Paula Clark, Process & Technology, Ciba Specialty Chemicals Corporation, 100 Ciba Road, McIntosh, AL 36553

The Pilot Plant fills the technology gap between a successfully demonstrated research concept and its practical implementation into a chemical facility. The Pilot Plant is responsible for demonstrating new technologies, creating design information for new processes, assisting in developing safe operating parameters, preparing customer samples for continuing product research, and production of small scale products. In order to achieve these objectives, identification, prioritization, and execution of several factors are necessary:

- Business Alignment - Hazards Assessment - Environmental Impact - Preventative/ Predictive Maintenance - Flexibility of Equipment - Efficient Turnarounds - Automation/Analytic Requirements - Adequate Resources

Once all elements have been thoroughly reviewed and implemented, the Pilot forms a “bridge” between research and production by providing the safest, most efficient technology for a successful product introduction.

572d Scaleup from Lab to Commercial in One Easy Step?

Mike J. Doll, HODer, Shell Global Solutions, 3333 Highway 6 South, Houston, TX 77082 and Jonathan H. Worstell, HODER, Shell Global Solutions US, 3333 Highway 6 South, Room E-2061, Houston, TX 77082.

Shell wanted to move new technology into commercially operating units. No pilot facilities existed for incrementally testing and proving the technology, so the decision was made to use one of the existing commercial units as a "big pilot plant". What were the obstacles encountered and overcome? Was the test successful? Did our reputations survive? This paper will describe how this plant test was proposed, managed, completed and the technology ultimately adopted as normal operating practice.

How to Design and Construct a Pilot Plant to Provide Maximum Value

Michael J. Zogg, Jr., David A Eversdyk, Daniel C. Floyd, & Pradeep Jain
The Dow Chemical Company, Freeport, TX, USA

Introduction

Today process engineers take advantage of economies of scale like never before. New fabrication and manufacturing techniques coupled with powerful computing technologies allow process equipment such as reactor vessels, heat exchangers, distillation columns, and processing pumps to be built to unprecedented sizes. The term "mega-train" has become synonymous with reactors that produce hundreds of millions of pounds of product per year. The Dow Chemical Company currently builds mega-trains for its solution polyethylene process that produce more than 300 kilotons per year of polyethylene resins. The so called mega-train produces product at minimum conversion costs while meeting product demand from customers located all over the world. In some cases, mega-trains are built to replace several older, smaller scale, and obsolete reactors that no longer compete on a cost-to-produce basis. The net result reduces the number of trains in operation while increasing overall production capacity and efficiency.

Given the impact to its "on-line" time, value is quickly lost when a mega-train is used for process or product development. Generally, new processes or products must be well defined and fully developed prior to implementation at this commercial scale. One can imagine the profit loss if a new unit operation is installed at this scale without proper research and development (R&D). Furthermore, it is easily envisaged that a new product with a market volume of 30,000 pounds (lbs) should not be produced on a reactor that operates at a production rate in excess of 75,000 pounds per hour (lbs/ hr).

Although name recognition and being the first to introduce a product to the market are advantageous, usually a company's market share is obtained and maintained through superior product performance. Over time and given that there is sufficient value, competitors often introduce similar products to the market and compete for that volume. To retain and even expand the volume of sales, the incumbent company must produce their product at a lower cost thereby enabling them to sell the product at a lower price and still retain profitability. Another alternative is to develop new and differentiated products to supplant the existing products while outperforming competitive products. Either approach requires research and development. As a result, R&D is integral to optimizing existing processes, developing new processes that yield lower operating costs, and developing new product families while supporting existing product lines.

Enter the pilot plant! As expected, coupling the arrival of the mega-train with an intensely competitive market place and competitors who can rapidly match even the best new products relatively quickly further elevates the importance of maintaining a "high performance" R&D organization. A key tool for such an organization is the pilot scale plant. A pilot plant that effectively replicates the manufacturing unit's capabilities can be used by the R&D team to properly support that manufacturing facility through process research and development efforts. Not limited to process support alone, a pilot plant can also be used to support the business's product development objectives. However, given the substantial differences in scale between

TM Trademark of The Dow Chemical Company

commercial and pilot plants, the detailed design and construction of the pilot plant is crucial. No design detail is insignificant and every difference in the way the two systems are designed or operated must be understood.

Not every pilot system has to be capable of supporting the entire R&D effort and prudent use of financial resources requires predetermining the purpose for the pilot plant. There is great utility for a piloting system to focus on one key aspect of research and development. As an alternative scenario to the pilot plant that fully mimics the production plant, researchers should consider a pilot system that is designed for a specific purpose such as testing new catalyst components that are developed as suitable replacements for existing components, testing new and different catalyst technologies for potential application in new product development, or acquiring data for kinetic modeling validation. These types of research efforts usually require that key catalyst activity data is collected; little product is required and process operability is generally not a concern. The critical data acquired through such efforts indicates if the catalyst is active or reaches a minimum acceptable yield. Although valuable, such reactor systems are clearly neither capable nor intended to be used for process support or market development. As a result these reactor systems can be very small, may even include the use of robot-automated micro-reactor systems, and are easily and quickly assembled at far less expense than fully functional pilot plants. The Dow Chemical Company strategically uses these types of assets as catalyst screening tools. They are highly efficient and enable hundreds of catalyst components and mixtures to be quickly tested for polymer forming activity. They use extremely small amounts of chemicals and minimize the preparatory work. Positive results can be studied in detail while non-active systems are discarded with minimal work. This tool has been successfully used to discover catalyst technologies that (after complete piloting) led to the successful commercialization of new product platforms such as VERSIFY™ Plastomers & Elastomers and INFUSE™ Resins.

Another kind of pilot facility is the small, inexpensive pilot system focused on one or two key aspects of research, thereby increasing the efficiency of the program and speeding up the overall research effort. As an example, The Dow Chemical Company conducted preliminary process evaluations for a monomer process development project in a 1 liter reaction system. Although, the 1 liter reactor was not representative of the pilot or the commercial designs and was not capable of delivering scale-up data, it was still valuable. Key screening and feasibility experiments were completed much more quickly at this scale and the experimental results facilitated the decision as to what should be studied in greater detail using the two orders of magnitude greater pilot plant. Preliminary kinetics and relative product effects were successfully collected. Research at the larger pilot scale was used to validate the 1 liter results while simultaneously evaluating operability and acquiring scale-up data for a commercial size reactor. Samples collected at the pilot plant were sufficiently large that full product characterization and performance evaluations could be completed. This scale was used to identify process issues, for example mixing problems or thermodynamic phase behavior that may result in precipitation (fouling) or boiling. Such problems often lead to issues with heat transfer, fluid flow, or product quality.

The Dow Chemical Company has constructed numerous pilot scale facilities for its polyethylene business. Each facility focuses on key research aspects and the facilities are strategically located around the world to ensure proper R&D support for the entire global business. The value these facilities bring to the company far outweigh the capital investment

™ Trademark of The Dow Chemical Company

and annual operating costs. The commercial success of existing product lines, like DOWLEX™ Polyethylene Resins and ATTANE™ Ultra Low-Density Linear Polyethylene Copolymer, coupled with the immensely successful launch of new product lines such as ENGAGE™ Polyolefin Elastomer, AFFINITY™ Polyolefin Plastomers, ELITE™ Enhanced Polyethylene Resins, VERSIFY™ Plastomers & Elastomers, and INFUSE™ Resins are directly linked to R&D piloting.

A creative R&D organization is critical to the long term viability of a company. As a result, numerous research and development opportunities await the engineer, including pilot plant design, construction, and operation. The authors intend to elucidate some of these opportunities by describing critical disciplines focused on the design and construction of a pilot plant. The authors anticipate that this will provide the reader with valuable insight into careers that can be developed through pilot scale research and development efforts.

Technical Design Requirements

It might surprise one to read that it can be more technically challenging to design a pilot scale reactor system than a full scale commercial scale facility. Proper business discipline dictates that the pilot plant is operational and critical technical design questions are answered prior to completing the commercial design. As a result, the pilot plant is constructed with a relatively high level of uncertainty. Due to the uncertainty of the process design, the engineering team necessarily has to incorporate extra piping and equipment (i.e. install multiple feeds, bypasses, extra purification, different types of heat exchangers in parallel, extra and/ or different pumps, redundant systems, significant instrumentation etc.) into the pilot scale reactor design. However, this strategy allows for the evaluation, modification, and optimization of the pilot scale process while completely defining the technology package prior to large scale construction.

The Dow Chemical Company has technology teams for many of its key product lines including polyethylene, polypropylene, glycols, epoxy resins, etc. People within these teams are subject matter experts regarding their overall process technology. The R&D team, including the pilot plant design and operating staff, are part of the core technology team for the business. In addition, The Dow Chemical Company has individuals that specialize in the design, construction, and operation of various unit operations (i.e. agitation systems, heat exchangers, distillation columns, etc.). As well, there are teams that specialize in areas including safety and environmental aspects of plant design and operation, as well as design ergonomics, process control, emissions control, state and federal emissions monitoring requirements, and compliance with worldwide food and drug application regulations. These subject matter experts work with the technical design teams (usually the pilot plant engineer) to address the important but ancillary design aspects for the facility. To properly leverage these skill sets, these experts may be involved with several core technology teams within the company. This networking process connects the general engineers and the specialists, field engineers and the theorists, the experimentalists, and the modelers, resulting in efficient team work and maximum value creation for the company. Other companies might employ a similar strategy or some skill sets might be directly incorporated into the pilot plant engineer's job requirements.

™ Trademark of The Dow Chemical Company

Typically, members of the core technology team develop a preliminary process flow diagram in an ideation session. The process flow diagram often includes ideal throughputs and production rates. As subject matter experts for the core technology, this team is most knowledgeable of the appropriate size at which to acquire data for scale-up. Although a production rate of 100 lbs/ hr might be sufficient for product development needs, the team might choose to design the pilot plant for a production rate of 1,000 lbs/ hr because the scale-up factor to the production plant is 10 times less for this design.

After generating the process flow diagram, a simulation software package is utilized to critique the process flow diagram and develop a detailed simulation of the pilot system with complete mass and energy balances. The Dow Chemical Company uses a simulation software package licensed from ASPEN Technology Inc. ASPEN is a process simulation tool used to scrutinize design scenarios and assist in developing the process configuration. Like all tools it has strengths and weaknesses. ASPEN simulation software's strengths include its ability to accurately calculate overall mass and energy balances, stream composition, heat exchanger duties, separation designs, and mechanical power requirements. However, it can not be used to calculate the detailed mechanical design of heat exchangers, reactors, distillation columns, pumps, etc. and is rarely used to generate accurate capital estimates. Other (often proprietary) tools are used for mechanical design of major equipment and to estimate the capital requirement. The aforementioned subject matter experts within The Dow Chemical Company have developed tools to facilitate the design of numerous unit operations such as mechanical agitation systems, heat exchangers, reactor vessels and distillation columns. Additionally, computational fluid dynamics is a tool that is used to determine the expected performance of various unit operations through computer simulation. When computational fluid dynamics is coupled with reaction kinetics and advanced phase equilibria models such as PC-SAFT[1] (perturbed chain, statistical associating fluid thermodynamics), the result is a powerful simulation tool for the pilot plant engineer.

After developing a process flow sheet with mass and energy balances, the core team uses its engineering judgment, knowledge of fundamental scientific principles, and experience to critique the proposed design. Even the best simulation tools will not help to determine the propensity of a heat exchanger to foul; the quality and effects of impurities in the raw materials; unsteady state effects during startups, reactor upsets, and shutdowns; what material to utilize in a packed bed; how to inject fresh feed into a reactor; or how to design a polymer handling system to minimize the generation and subsequent contamination of product due to cross-linked polyethylene (i.e. gels). The following examples show the impact that good engineering judgment has on pilot plant design and how important it is to understand the similarities and differences between the pilot design and the commercial design.

As mentioned earlier, The Dow Chemical Company's polyethylene business has several pilot scale facilities that the R&D team uses to support the business. One key design difference between some pilot and manufacturing plants is the lack of a pilot scale monomer recycle system. The mass of unreacted monomer exiting the pilot plant reactor system is relatively small and the team chose not to install monomer recycle systems for these facilities. Instead this stream, also containing unreacted hydrogen (used as a chain terminating agent), is burned as a waste stream using a thermal oxidizer. However, the total concentration of hydrogen fed to the reactor is manipulated to control the molecular weight of the product. The R&D team invested significant time studying the impact of not recycling the unreacted monomer stream and the resulting capability differences between the pilot and the manufacturing plants. The team concluded the key difference was the ability of the pilot scale reactor system to operate at lower hydrogen levels in the reactor feed due to the lack of a monomer recycle system. As a result, for a given set of reaction parameters, the R&D team was capable of making higher molecular weight products. After quantifying this difference and designing reactor conditions to maintain a level operating field, the pilot reactor system was successfully used for scale-up to the manufacturing plant.

Another example of how critical it is to define capability and operating discipline differences early in the R&D effort relates to the replacement of a new cocatalyst component. To maximize flexibility, one of The Dow Chemical Company's pilot facilities was constructed with 4 extra vessels (with feed pumps) to be used for catalyst component storage and feed to the reactor. Additionally, the pilot plant was constructed with 3 catalyst feed injectors into the reactor. This flexibility allowed the R&D team to study new catalyst technologies, often without making physical modifications to the reactor system. Obviously, a manufacturing plant is constructed with only strategic sparing of equipment. In particular this production plant was not constructed with multiple catalyst injectors. Instead, all of the catalyst components were injected through a single injector. Normal operating discipline required the pilot plant team to inject fluid through each of the injectors to prevent fluid back flow into the injector and injector plugging. The R&D team had proven there was no discernable operability or resultant product characterization difference between the way the pilot plant injected the traditional catalyst package (separately injecting the components into the reactor) and the manner in which the manufacturing unit injected the catalyst components (as a combined mixture into the reactor through a single injector). When the replacement cocatalyst study was initiated, the team designed experiments to determine if there was any potential downside to the design and operating discipline differences. After successfully using the new cocatalyst with typical pilot plant operating discipline, the team modified the injection strategy to mimic the production plant design. Results were not very pleasant! Almost immediately the injector plugged. Subsequent laboratory studies showed that the activated catalyst, when using this cocatalyst, formed an insoluble precipitate that fell out of solution with the carrier solvent and plugged the injector. As bad as this scenario was, the pilot plant was exactly the correct place to identify the problem! Prior to commercial implementation, the R&D team worked with manufacturing personnel, disclosing this new information and defining an implementation strategy including the required physical modifications to the catalyst injector.

The following is an example of how a new process flow diagram and subsequent plant design was generated from an ideation session after the discovery of a new and intriguing catalyst technology. The Dow Chemical Company had used a high temperature catalyst technology with an adiabatic reactor system to make DOWLEX™ Polyethylene Resins. However, with the advent of low temperature polymerization catalysts that yielded differentiated products, the core technology team realized that a new reactor technology was required. The adiabatic reactor system that had been successfully employed for DOWLEX™ Polyethylene Resins production was a poor fit with the new catalyst technologies. In the adiabatic solution polyethylene process, the heat of polymerization is removed using the heat capacity of the solvent. To maintain high production rate at low temperature requires more solvent dilution, thus significantly increasing costs to cool, transport, purify, and store the recycle solvent stream. Additionally, the activity of the new catalyst technology is negatively

™ Trademark of The Dow Chemical Company

impacted due to the dilute environment. In other words, adiabatic reaction conditions destroyed the advantages that the new catalyst technologies offered. As a result, the R&D team designed and constructed a novel, patented "next generation" reactor technology for solution polyethylene, the non-adiabatic reactor. Critical items the team had to address included how to properly design the reactor's heat exchanger, how to remove the heat of reaction without fouling the heat exchanger or other parts of the reactor system, and how to mix low viscosity feed streams into a high viscosity reaction mixture. Such issues are quite challenging to the engineer since diffusion is the dominant mode of mass and energy transfer within the reaction fluid and the highly viscous fluid flow is in the laminar flow regime. Diffusion limits may constrain the reaction kinetics or phase equilibria. Significant research and development was invested to design an optimal heat exchanger for most efficient heat transfer, with minimal capital cost or operational fouling. Additionally, R&D efforts defined the most efficient manner in which to inject and quickly mix cold, low viscosity feed into a comparatively hot, high viscosity, fast reacting solution. The results of such process developments enabled the successful launch of various product families such as AFFINITY™ Polyolefin Plastomers, ENGAGE™ Polyolefin Elastomer, VERSIFY™ Plastomers & Elastomers, and INFUSE™ Resins at economic costs with excellent product quality.

A critical product defect in the polyethylene business is gels in the final product. Gels can be formed by a variety of mechanisms. Years of engineering experience proves valuable when designing a pilot plant system for minimum contamination due to polyethylene gels. For example, the core polyethylene technology team knows that every measure must be taken to exclude impurities from the reactor and separation systems. Additionally, it is understood that the pipe and equipment volume in high heat areas (i.e. the solvent/ polymer separation system) must be minimized and dead spots eliminated. As a result, this section of plant design has been streamlined to allow quick and effective processing of product as well as cleaning of the system. Employing these disciplines and many other preventative measures in the design affords the team every opportunity to manufacture product (at any scale) with minimal gel contamination.

When designing a reactor system, engineering fundamental principles must always be utilized to define the critical parameters. These parameters will be evaluated during pilot plant operation and the acquired data will be used to define the final technology package. Use caution when defining the critical scale parameters as they are not always apparent. Dimensionless numbers such as the Reynolds number, Prandtl number, Peclet number, Stanton number, Nusselt number etc. are just a few of the traditional scaling parameters. When designing a first generation reactor system, traditional design strategies must be employed because there is limited engineering experience. However, once a reactor system is constructed and functional, it is the research engineer's responsibility to study the reactor system and succinctly define the key design parameters that will impact scale up.

The following is an example of non-traditional scaling. During pilot scale work the process R&D team discovered that a slurry additive did not remain suspended in solvent when it was fed to the post reactor system unless the fluid velocity in the pipe exceeded a critical value. Below this critical velocity, the additive particles fell out of suspension, agglomerated at flow disturbance areas (i.e. elbows and valves) and plugged the pipe. This was a critical piece of information that had to be defined in the technology package. It was critical that the team

™ Trademark of The Dow Chemical Company

document that the key parameter when designing the pipe size for feeding additive was not the Reynolds number, but instead the fluid velocity.

Safety Reviews and Engineering

Technical process design is only the start of the overall pilot plant design. It is equally important to incorporate safety features into the plant design that allow for safe operation, even when human interaction might inadvertently or unknowingly try to force the process technology into an unsafe operating regime. Like all companies in the industry, The Dow Chemical Company has a rigorous safety program and requires the technical design team to conduct numerous safety reviews for the technical design package. Generally, the pilot plant engineer leads the safety reviews, generating plausible "worst case" scenarios and working with the particular safety subject matter experts to define appropriate resolutions. Safety design reviews include but are not limited to company developed evaluations such as Fire and Explosive Index (F&EI) and Chemical Exposure Index (CEI) calculations as well as reactive chemicals testing and chemical compatibility reviews, Hazards of Operability (HAZOP) studies, ergonomics reviews, Loss Prevention Principles (LPP) reviews, equipment overpressure relief design calculations, proper design, location, and installation of safety showers and emergency exits, and the establishment of personal protective equipment (PPE) requirements. Many of the reviews (i.e. F&EI and CEI) are used to evaluate the exposure risk to personnel in nearby control rooms and offices. As part of the technology package all hazards and controls must be reviewed and approved by the technical design team and the process safety team. Many of these safety programs have been developed by The Dow Chemical Company and shared with the global chemical industry over the years. Criteria always conform to regulations by government agencies, and are often more stringent than the standards required by these agencies.

Automation

Today, computing power is relatively inexpensive and process control systems permit process control and automation to be fully exploited. The positive impact on the industry is undeniable! Automation has made processes more efficient, safer, and resulted in capabilities to repeatedly manufacture products with high quality. Other advantages include less human interaction and subsequently less exposure and more efficient interaction. Another benefit is the ability to collect and store large amounts of process data. Automation is also mistake-free: a computer will not miss the timing of a critical step in a sequence, accidentally transcribe the value when entering a set point, etc. The following two examples will further demonstrate how remote operation has benefited the industry. Consider the start-up of a multi-stage reciprocating compressor used to compress a hydrocarbon gas stream to a few thousand atmospheres in pressure. Obviously, the ability to start the compressor from a control room located at a safe distance virtually eliminates concerns regarding human safety. Another example is the situation in which a batch of catalyst components must be mixed four times an hour, for at least 10 minutes prior to injection into a continuously running reactor. Without process control and automation, a single pilot plant operator would dedicate their entire shift to this single task.

Designs to Protect the Environment

TM Trademark of The Dow Chemical Company

A pilot plant design must include means by which to protect the environment. All waste streams must be minimized and properly disposed of. Innovative ways to reduce the waste generated from a process and improved destruction techniques are critical to the long term viability of a technology. Gas waste is generally burned via flares or thermal oxidizers. These unit operations must be designed to adequately accommodate normal operation as well as unsteady state operation and surges of venting so that optimum destruction efficiency is maintained and greenhouse gas emissions are minimized. Waste must be properly removed from water and solids before discharging to the environment or land filling. To protect wildlife and the surrounding environment, the facility must be designed to contain solid or liquid chemicals in the event of a spill. Additionally hydrocarbon gases released from vessels through over-pressure protection devices (i.e. pressure safety valves) generally must relieve to a thermal destruction unit operation. Even rainwater run-off from the pilot plant concrete slab must be collected (generally in a sump) and sampled for organic content prior to untreated discharge to the environment. In today's world, designing a plant (regardless of scale) to minimize and properly handle waste is a critical design element.

Construction, Ergonomics, and Commissioning

It is normal for the construction team to uncover minor obstacles that are difficult to discern from a piping and instrument diagram (P&ID) or isometric drawing. An experienced construction foreman will bring these issues to the attention of the pilot plant engineer. Typical issues include relocating valves to more user friendly locations, designing a new pipe lay-out around a piece of equipment for improved access by operations and maintenance, and relocating or adding additional valves to be used as vents and drains. It is often critical for personnel to be able to gain safe and ergonomically sound access to equipment within the facility for safe operation or maintenance. Like safety incidents, poor ergonomics can lead to serious injuries.

The commissioning team has the task of thoroughly inspecting the pilot plant to verify that construction matches the design, the reactor system is leak free at operating conditions, and the process control program performs as designed under normal and emergency operation. Even at the pilot scale these tasks may take many weeks to complete. It is the responsibility of the pilot plant engineer to develop a detailed, easy to use plan that guarantees every line, instrument, and piece of equipment is inspected and verified for accuracy.

Initially, the commissioning team inspects the plant, comparing each line on the P&ID to the final product in the field. The team must review each pipe to ensure it was fabricated using the correct material of construction, the correct pipe-class, the pipe has the correct surface treatment inside and out (i.e. galvanized pipe, no paint, paint for cold temperature service, paint for high temperature service, acid washed, etc.), is correctly insulated where required, has the correct gasket type and material of construction for flanged connections, the correct sealant for threaded connections, the correct bolts for mating the flanges, and adequate bracing and support. Even details such as appropriate code stamps on welds must be verified. A similar analysis of all equipment, valves, and instruments in the field must be completed in addition to verifying that the installation location of each relative to the other matches the corresponding P&ID.

TM Trademark of The Dow Chemical Company

The team must also verify that every instrument is installed in the correct location and meets the instrument design specifications. In addition, it must be verified that every instrument is wired to the process control system correctly and moves/ measures its respective parameter correctly and the instrument range matches that on the control system. Prior to utilizing rotating equipment each motor must be tested (while uncoupled) for proper ground and rotation. After the motor rotation is verified as correct the equipment is coupled to the motor and aligned within the equipment specifications.

Next, the team must verify the mechanical integrity of the system. This is initially completed using a nontoxic chemical (i.e. water, air, or nitrogen) and then verified with the process fluid. New utility systems are commissioned and placed in operation first. Upon completion of commissioning the utility systems, the team will start up process systems in a predetermined logical and sequential manner, testing program logic throughout. It is quite common to install filters in the suction of all rotating equipment to protect the equipment from construction debris. Typically the filters are used to protect the equipment from dirt, scale, and sand. However nuts, bolts and gloves have managed to get into process piping during construction. After the system has been flushed, the filters are removed and the facility is ready for start-up and normal operation.

Conclusion

A successful piloting program requires the technology team to define the number of pilot plants as well as the required utility for each. Furthermore, complete R&D capability requires the ability to replicate the manufacturing facility's capabilities in at least one of the pilot systems. This strategy allows the R&D team to develop the technology package for the commercial scale and provide support once the commercial facility is operational. As a result, the technical design of a pilot plant incorporates numerous engineering disciplines and skill sets. However, required skill sets were illustrated throughout the paper that show technical design is only one facet of the overall pilot facility design. Numerous other skill sets are required to construct a facility that delivers quality research while operating in a safe and environmentally conscientious manner.

Reference

1. Tumaka, F.; Gross, J.; Sadowski, G., Thermodynamic modeling of complex systems using PC-Saft. Fluid Phase Equilibria, 2005. 228-229: p. 89-98.

572f Thermal Cracking Studies By Design Of A Pilot Plant

Fatemeh Abniki, Process Eng.Dept. of National Petrochemical Company,Ghadir Group, Tehran, Iran and Ehsan Bakhshi, R&D Center of National Petrochemical Company, Tehran, Iran.

Thermal cracking of hydrocarbons from ethane up to gasoil is one of the most important processes for production of olefins, the basic feedstock for the petrochemical industry. The key to successful process control in the manufacture of ethylene, is temperature control of thermal cracking furnaces. This paper studies a computerized thermal cracking pilot plant and its capabilities for experimental activities. A computer program was developed for the monitoring and control purposes. The effect of temperature on ethylene yield is investigated by simulation and experimental studies. The results show good agreement between. Comparison of these results with industrial data, show that, they have the same trends. Coil Outlet Temperature (COT) is an important parameter affecting yield of ethylene production and therefore it should be controlled. Since furnaces are the first step in the production process, disturbances that occur due to the furnace operation will effect the entire process. COT is controlled by measuring the cracked gas temperature in the coil outlet and manipulating the heat input to the furnace. This loop is one of the most important loops in the control of thermal cracking furnaces. Fluctuations on COT is caused by the coke deposited on the COT measuring thermocouple. Furnace temperature tracks the furnace set point. Increasing the oscillation at higher set points, are due to nonlinear plant dynamic and using constant controller tunings. Based on the kinetic model a computer program for simulating the plant is developed. The software program is used to investigate the effects of temperature on the product yields. Simulation results indicate that increasing COT, increases the ethylene yield. Simulation results was tested experimentally by the pilot. Comparison of simulation, experimental and industrial results show, those have the same trends. The discrepancies between industrial and simulation results refer to the reactor configuration and the discrepancies between simulation and experimental results are due to the nonlinear and unknown furnace dynamic model, and some measurement errors in experiments.

Web Page: www.nipc.net

Critical Process Parameter Determination for a Sterile Liquid Filled Syringe Product

An important consideration in the technology transfer of a pharmaceutical finished product into manufacturing is the determination of critical process parameters (CPPs). After the overall process flow is devised, the processing conditions for each individual unit operation can be determined.

A case study of a new manufacturing suite for a sterile liquid product is used as an example. The particular unit operations are buffer preparation, tangential flow filtration (i.e. cross flow filtration) for the buffer and active ingredient, final batch compounding (non-sterile), sterile filtration, and syringe filling in an isolator. Process parameters, equipment capabilities, predicted operating ranges, critical operating limits, and the impact of exceeding critical limits are considered for each unit operation. This analysis demonstrates an approach to the process review and assessment of the key process parameters for a new commercial sterile liquid filled syringe protein product.

648a Troubleshooting and Performance Recovery of a Pharmaceutical Manufacturing Process Involving a Novel Asymmetric Hydrogenation and a Chiral Crystallization

Osama Sudah[1], Aaron Moment[1], Carlos Burgos[1], Cindy Starbuck[1], Andrew Clausen[1], Mahmoud Kaba[2], Carlos Sanchez[1], Ryan Adorno[1], Mabel Suarez[1], Alvin Negron[1], Luis Sierra[1], and Ramon Huertas[1]. (1) Chemical Process Development & Commercialization, Merck & Co., Inc., RY818-C314, P. O Box 2000, Rahway, NJ 07065, (2) Merck & Co. Inc, Barceloneta, PR

A novel pharmaceutical manufacturing process to produce a chiral intermediate was successfully scaled-up to 2000 gallons in the factory during technology transfer. The manufacturing process involved several unit operations, and featured a unique high pressure enantioselective hydrogenation to produce the chiral intermediate at > 93% ee (enantiomeric excess), as well as a distillative chiral crystallization to purify the product to > 99% ee.

During the second manufacturing campaign, erosion in chiral purity, and elevated impurity levels were observed in the isolated intermediate. This erosion in purity was unexpected, and had potential implications on the quality of the product in the subsequent step. As a result, a thorough investigation was launched during the campaign to uncover the root causes of the process deviation, and to promptly recover batch performance.

The troubleshooting effort took into account the differences in operational parameters for each unit operation between campaign I and II, as well as process optimization changes that were in effect for campaign II. The confounding nature of the process parameters posed a challenge to the investigation and made a deviation analysis particularly difficult. Nevertheless, the investigation successfully identified two key factors contributing to the erosion in purity: the operating pressure during the distillative chiral crystallization and the overall train cycle time. Consequently, a process recovery strategy was implemented, and batch performance was restored for the remainder of the campaign with safeguards put in place to protect against future purity erosion. The troubleshooting approach, along with its findings, and highlights of the process recovery strategy will be presented.

648b Control of a Membrane Separation Process by UV / Vis P.a.T

James Marek, Eric Stoner, and James Schilling. GPRD Process R&D, Abbott Labs, 1401 Sheridan Road, R13-3, North Chicago, IL 60064

This project demonstrated the benefits of controlling an Ultrafiltration/Diafiltration (UF/DF) membrane separation process by UV/Vis photometry. API freebase concentrations in the retentate (product) and permeate (waste) streams were monitored continuously. The ExP unit performed robustly and provided useful data in an engineering run, three ensuing validation runs, and a fifth manufacturing run. The UV/Vis PAT could streamline operation of the membrane process, potentially eliminating four in-process analytical samples and an additional 36 permeate grab samples taken to verify membrane integrity and detect product losses. The features of the UV/Vis photometer will be described, as will the design of the in situ sample loops, including full color photographs of the field installation and data collected.

648c Expedite Development and Commercialization of Pharmaceuticals with Process Modeling and Scheduling Tools

Demetri Petrides and Charles Siletti. INTELLIGEN, Inc., 2326 Morse Avenue, Scotch Plains, NJ 07076

The successful development, scale-up, and commercialization of active pharmaceutical ingredients (APIs) and final pharmaceutical products is a challenging task that requires collaboration of professionals from many disciplines. Process simulation and scheduling tools can facilitate this task by introducing a common language of communication among the various teams involved in process development and product commercialization. They can facilitate answers to the following questions: What is the impact of alternative synthesis routes on the economics and environmental impact of a process? What are the cost sensitive steps of a process that need further development and optimization? What process changes can reduce cycles times and lower manufacturing cost? What changes are required in an existing multi-purpose facility to accommodate the manufacturing of a new product? Our experience in addressing the above and other pertinent questions will be presented using examples from the pharmaceutical and biopharmaceutical industries. The examples will cover production of bulk intermediates as well as manufacturing of final pharmaceutical products.

Web Page: www.intelligen.com/literature

648d Modeling Based Approach towards on-Scale Implementation of a Methanethiol Emitting Reaction

Brenda Remy, Shawn Brueggemeier, Olav Lyngberg, D Lin, Alex Marchut, and L Hobson. Bristol Myers Squibb, New Brunswick, NJ 08902

Reactions emitting volatile organic compounds (VOC) pose unique environmental challenges which must be addressed upon scale-up. Efficient implementation of such processes requires thorough understanding of the VOC generation rate as well as knowledge of the ultimate fate of the VOC. Modeling tools can assist in developing this fundamental process understanding and, in doing so, may reduce the time required for scale-up. Using DynoChem, a kinetic model of a methanethiol generating reaction was developed. This model was then used to optimize the reaction temperature profile, with the goal of evenly distributing the methanethiol off-gas flow rate throughout the course of the reaction. Through the use of modeling, two reaction temperature profiles were identified that reduced the maximum off-gas flow rate by 75 % while only increasing the reaction cycle time by 15 %. The two temperature profiles were verified via lab scale experiments and then successfully implemented at the pilot plant scale. This modeling based approach proved effective at quickly identifying and implementing process solutions to mitigate the environmental risk associated with scale-up of this methanethiol producing reaction.

648e Applications of Mass Spectrometry for Monitoring Gaseous Streams during Processing

John P. Gardner[1], Jacob R. Remacle[1], Steven J. Doherty[2], Daniel J. Creeden[2], Kevin D. Seibert[3], Zhonghui Wan[4], Amanda L. Mc Daniel[4], John P. Schafer[1], and Gary A. Richardson[1]. (1) API Operations Engineering, Eli Lilly and Company, Lilly Corporate Center, Indianapolis, IN 46285-4816, (2) PAT, Eli Lilly and Company, Lilly Corporate Center, Indianapolis, IN 46285-4816, (3) Development Engineering, Eli Lilly and Company, Lilly Corporate Center, Indianapolis, IN 46285-4816, (4) Chemical Product Research and Development, Eli Lilly and Company, Lilly Corporate Center, Indianapolis, IN 46285-4816

In the course of manufacturing an Active Pharmaceutical Ingredient (API) to supply material to support toxicology studies and clinical trials, it is often desirable to monitor by-products of the chemical processes. Mass Spectrometry (Mass Spec) is a reliable method for tracking the gaseous by-products of the reactions or vapor emissions during drying procedures. Additionally, it is useful for determining the efficiency of a scrubber or for optimizing the spin time during centrifugation operations.

The authors will present a comparison of results from laboratory experiments with Pilot Plant campaigns and discuss equipment configuration for monitoring gas streams. Example applications include monitoring corrosive by-products from an acyl chloride formation, insights into the mechanism and scrubber efficiency of an oxidation reaction and determination of the drying cycle during centrifugation of an API.

A Quality by Design (QbD) Approach for Pharmaceutical Capsule Filling Unit Operation: Identifying Critical Process Variables and Detecting Possible Interactions

Huiquan Wu[1a], Lin Xie[2], Meiyu Shen[3], Mansoor A. Khan[1], Robbe C. Lyon[1], Larry Augsburger[2], Stephen W. Hoag[2]

[1]Division of Product Quality Research (HFD-940), OTR/OPS/CDER/FDA, Life Science Building 64, 10903 New Hampshire Ave, Silver Spring, MD 20993
[2]School of Pharmacy, University of Maryland, 20 N. Pine St., Baltimore, MD 21201
[3]Division of Biometrics VI, OB/OTS/CDER/FDA, Building 22, 10903 New Hampshire Ave, Silver Spring, MD 20993
[a] Corresponding author; Dr. Huiquan Wu. Email: huiquan.wu@fda.hhs.gov; Phone: 301-796-0022

Abstract: Capsule filling process performance is dependent on capsule filling machine and the formulation. Variation in capsule weight clearly results in variation in drug content. The purpose of this study was to use an integrated QbD approach to identify critical process variables and detect possible interactions between variables for a pharmaceutical capsule filling unit operation. The formulation consisted of 4% aspirin and 96% MCC. Two full factorial designs were created to include selected formulation [aspirin type (coarse and fine), filler MCC (Avicel® PH 200 and PH 301), lubricant level (0 and 0.5%)], and process [(capsule size (#0 and #1), filling machine type (Zanasi and H&K)] variables. A total of 24 batches were manufactured. Ten capsule samples were collected at each of the predefined time points throughout each production run. Particle size, flowability, bulk density and tapped density were measured. The coefficient of variation (CV) for capsule weight and aspirin content were determined for each interval sampled (CV_i). Main effects and interactions were assessed by ANOVA. Principal component analysis (PCA) was used to analyze the variability in the powder's physical properties. It was found that for average capsule fill weight, machine type was a significant variable. For percent aspirin content, capsule filling machine type, lubricant level, interactions between filler type/capsule size and lubricant level/capsule size were found to be significant. For CV_i of aspirin content, only aspirin particle size was significant. For capsule weight average CV_i, filler type, lubricant level, capsule filling machine type, and interaction between aspirin particle size/capsule filling machine type were found to be significant. Carr's compressibility index was found to be the main contributor to the first principal component (PC) in the dataset of the powder physical properties. This work demonstrated the possibility of integrating DOE, multivariate statistical data analysis, and product/process characterization for identifying critical process/product variables and detect possible interactions between the formulation/process variables. The knowledge gained through this integrated QbD approach may provide scientific justifications for certain regulatory actions in the section of Chemistry, Manufacturing, and Control (CMC).

Key words: Quality by Design (QbD), design of experiments, multivariate statistical data analysis, product characterization, capsule filling process characterization, capsule filling weight variation, content uniformity, SUPAC, level of change.

INTRODUCTION

Recent FDA Guidances[1-2] (PAT, Quality System, etc.) and ICH guidances[3-4] (Q8 and Q9) recognized the importance of using a Quality-by-Design (QbD) approach for pharmaceutical development, process and product design, and manufacturing improvement. Design of Experiment (DOE) as a key element of the QbD paradigm, has been widely adopted in other industry sectors for many applications including process robustness and product design[5], scale-up and optimization, and process control[6]. If DOE is used appropriately in pharmaceutical sector, in addition to the above-mentioned benefits, most importantly it can be an effective tool to mitigate risk of producing pharmaceutical products of low quality. DOE is an efficient tool to develop process understanding and to construct a design space. As such, the outcomes of DOE could provide scientific justification and rationale for risk-based pharmaceutical cGMP regulation in the 21st century.

The capsule filling process is one key unit operation for manufacturing pharmaceutical solid dosage form. Capsule filling process performance is dependent on the capsule filling machine and the formulation. Variation in capsule weight clearly results in variation in drug content. Normally there are two interrelated safety issues about the capsule product quality: capsule weight variation and content variation or content uniformity issue. A previous study[7] showed that appropriate selection of pin settings and bed height can reduce the weight variability seen, especially with poorly flowing high-dose formulations. Integrating of DOE, physical/chemical characterization, and multivariate data analysis (MVDA) may provide an integrated approach for understanding the relationship among formulation variables, process variables, and quality attributes of intermediate and final products[8].

The aim of this study was to integrate DOE and MVDA for identifying critical process variables and detecting possible interactions between variables for a pharmaceutical capsule filling unit operation. The considerations of using such an integrated approach for the risk-based pharmaceutical regulation in the 21st century were discussed.

MATERIALS AND METHODS

Experimental Design

The formulation comprised 4% aspirin and 96% MCC. Two full factorial designs were created to include 5 factors, which were selected formulation variables such as aspirin particle size (APS) (coarse and fine), type of filler MCC (FT) (Avicel® PH 200 and PH 301), and lubricant level (LL) (0 and 0.5%), and process variables such as capsule size (CS) (#0 and #1) and filling machine type (MT) (Zanasi and H&K). For Zanasi machine, a 2^4 full factorial design was used. For H&H machine, a 2^3 full factorial design was used. A total of 24 batches were manufactured. Ten capsule samples were collected at each of the predefined time points throughout each production run. Particle size (Malvern Mastersizer), flowability (Shear Cell and a Flodex Tester), bulk density and tapped density (Scott Volumeter and the USP tapped density method) were measured.

Physical Characterization of Bulk Powder

Particle Size Analysis

Particle size was measured by laser diffraction with a Malvern Mastersizer (Malvern Inc., Worcestershire, UK). The dry powder feeder was operated at an air pressure 60 psi, and a sample size of 1.0 g to 1.5 g. The volume based particle size data mean particle size D[4,3] and mode diameter were used. A microscope (Nikon, Eclipse ME 600, Image Systems Inc., Columbia, MD) was used to take the pictures.

Density Measurements

True densities of raw materials were measured using a helium pycnometer (USP26/NF21 <699>) (Helium Pycnometer 1305, Micromeritics Instument, Norcross, Georgia). The true density of a blend was calculated from the formulation composition. The bulk densities were measured by using a Stampfvolumeter (USP26/NF21 <616>). The tapped densities of powder blends were measured using a STAV 2003 Tap density tester (USP26/NF21 <616>). Carr's compressibility index (CI) value was calculated using the formula[9]:

$$Carr's\ compressibility\ index = \left(\frac{Tapped\ density - Bulk\ density}{Tapped\ density}\right) * 100 \quad (1)$$

Generally a smaller Carr's compressibility index value indicates better powder flow.

Minimum Orifice Diameter

Minimum orifice diameter (MOD) was determined using a Flodex™ tester (Hanson Research Corporation, Chatsworth, CA). The sample size was 50 g. The smallest orifice that the bulk powder passed through and left an upside-down truncated cone instead of a cylinder shape rathole was used as the minimum orifice diameter. A smaller MOD value indicates better powder flow.

Shear Cell Analysis

An annular shear cell was used to determine the angle of friction (AF), angle of effective friction (AEF) and cohesion (Co). A powder bed height of 1 cm, a sample size of approximately 40 to 60 grams of bulk powder and a lid with baffles were used for all tests. The tests were carried out using a constant rotating speed of 8 rpm.

ASP Assay of Filled Capsules

All aspirin samples were placed in a 0.1 N NaOH solution and sonicated for 30 min to completely dissolve the aspirin. The suspension was centrifuged at 2000 rpm for 10 min; the supernatant was diluted with 0.1 N NaOH to suitable concentration for measurement by UV absorbance at 298 nm. The 0.1N NaOH solution was used as a control. The standard curve was linear (R^2 = 0.999).

Data Analysis

For a given quality attribute, the coefficient of variation (CV_i) at manufacturing time t_i was defined as the ratio of the standard deviation of samples collected at time t_i vs. the mean value of samples collected at time t_i. It was calculated by the following formula:

$$CV_i = \frac{(STD)_i}{(Mean)_i} \qquad (2)$$

The coefficient of variation (CV) for capsule weight and aspirin content were determined for each interval sampled (CV_i). Main effects and interactions were assessed by ANOVA using SAS 9.1.3 software program (SAS Institute Inc., Cary, NC). Principal component analysis (PCA) using Unscrambler 9.2 software (Camo Technologies, Woodbridge, NJ) was conducted to analyze the variability in the powder's physical properties.

RESULTS

1. Average capsule filling weight

Table 1 shows the ANOVA results for the average capsule filling weight. The following variables or two-way interactions were found to be statistically significant at the 0.05 confidence level: APS, FT, LL, CS, MT, FT*LL, FT*MT, LL*CS. These results are scientifically important as they could tell when any of these formulation or process variable changes, there is a statistically significant difference to be observed among the average capsule filling weight at the 0.05 confidence level.

Table 1. ANOVA results for average capsule filling weight

Source	DF	Mean Square	F Value	Pr>F
APS	1	301.11	5.13	0.0387
FT	1	13463.31	229.47	<.0001
LL	1	410.44	7.00	0.0184
CS	1	7446.40	126.91	<.0001

MT	1	1044.91	17.81	0.0007
FT*LL	1	1511.94	25.77	0.0001
FT*MT	1	4034.25	68.76	<.0001
LL*CS	1	499.23	8.51	0.0106

2. **Capsule weight average CV_i**

For the capsule weight average CV_i, only the factors FT, LL, MT, and interaction term APS*MT were found to be statistically significantly influential at the 0.05 confidence level.

3. **Average ASP content (weight)**

For average ASP content (weight), only the factors FT, CS, MT, and FT*LL were found to be statistically significantly influential at the 0.05 confidence level.

4. **CV_i for ASP content**

For CV_i for ASP content, only the factor APS was found to be statistically significantly influential at the 0.05 confidence level.

5. **PCA model for variability characterization on powder physical properties**

PC1 and PC2 explain 94% and 5% of the total variability, respectively. CI and MOD contributes the most to the PC1. FF contributes most to the PC2.

DISCUSSION ON USING INTEGRATED APPROACH FOR RISK-BASED PHARMACEUTICAL REGULATION IN THE 21ST CENTURY

"Critical control points" is one key term highlighted in FDA's recent initiative "Pharmaceutical cGMPs for the 21st Century: A Risk-Based Approach"[10]. Over the last few years, the FDA and the industry representatives regularly speak of identifying the "critical control points" for a given process, and the need to support the process with well-conceived design, control, and maintenance procedures. Key elements of a unit operation or process that require special attention are often identified through risk analysis tools such as Hazard Analysis and Critical Control Point (HACCP) and Failure Mode and Effect Analysis (FMEA). In present work, we demonstrated the possibility of integrating DOE, MVDA, Product and Process Characterization for identifying critical product variables, critical process variables, and detecting significant interactions between formulation and process variables. Thus, critical variables can be identified early at pharmaceutical development stage. Therefore, monitoring and controlling these critical factors are well justified and can be taken into design considerations for manufacturing.

In addition, identifying critical process variables and interactions through an integrated approach highlighted in the work can provide a rationale for some science-based regulatory actions in the area of Chemistry, Manufacturing, and Control (CMC). For example, certain post-approval changes proposed by the pharmaceutical companies must seek FDA's pre-approval before the companies could actually implement those changes in the process and manufacturing line. For this study, we demonstrated that:

(1) For average capsule fill weight, machine type was found to be a significant factor. This indicated that, for one fixed formulation, we may not switch capsule filling machines between a Zanasi machine and an H&K machine if we want to produce products with consistent capsule filling weight. According to the FDA's SUPAC Guidance[11], this kind of change may fall into the category of "Level 2 Changes" of Manufacturing, because the Zanasi and H&K machines have different designs for the manufacturing process.

(2) For percent aspirin content, capsule filling machine type, lubricant level, interactions between filler type/capsule size and lubricant level/capsule size were found to be significant. Therefore, changes on these variables may trigger regulatory actions, because the drug content is such an important quality specification that we have to meet. Any change in this regard would need regulatory scrutiny, as defined in the FDA's SUPAC Guidance. According to the

FDA's SUPAC Guidance, these changes may fall into the category of either "level 2 changes" of manufacturing or "level 2 changes" of components and composition.

Although the SUPAC Guidance does describe levels of change in the CMC area in a qualitative way, it does not describe how to classify these changes in a quantitative way. This work, while focusing on identifying critical variables and critical control points for the risk-based cGMPs for the 21st century, provides a case study of using an innovative integrated approach for quantifying change levels described in the FDA's SUPAC guidance.

CONCLUSIONS

Integration of design of experiment (DOE), multivariate statistical data analysis, and product/process characterization can provide a rationale approach to identify critical process/product variables and detect possible interactions between the formulation/process variables. This study indicated that capsule filling machine type was a significant variable for average capsule fill weight. For percent aspirin content, capsule filling machine type, lubricant level, interactions between filler type / capsule size and lubricant level / capsule size were significant. In addition, based on the ANOVA results of the DOEs, the level of change and its related regulatory submission requirements were discussed in the context of the FDA's SUPAC guidance. This study illustrated an innovative integrated approach for quantifying levels of change associated with SUPAC guidance, thus may help promote science-based pharmaceutical regulation.

ACKNOWLEDGMENT

This work was financially supported by FDA CDER RSR Grant 04-16. Dr. Ajaz S. Hussain, formerly at OPS/CDER/FDA, is acknowledged for technical discussion and support in the early stage of this project. Drs. Stella G. Machado and Yi Tsong at DBVI/OB/OTS/CDER/FDA are acknowledged for statistical guidance and support.

Disclaimer: *The views and opinions expressed in this paper are only of the authors, and do not necessarily reflect the views or policies of the FDA.*

REFERENCES

1. FDA (2004). Guidance for Industry, PAT — A Framework for Innovative Pharmaceutical Development, Manufacturing, and Quality Assurance. Available at: http://www/fda.gov/cder/guidance/6419fnl.pdf.

2. FDA (2006). Guidance for Industry, Quality Systems Approach to Pharmaceutical CGMP Regulations. Available at: http://www.fda.gov/cder/guidance/7260fnl.pdf.

3. FDA/ICH (2006). Guidance for Industry. Q8 Pharmaceutical Development. Available at: http://www.fda.gov/cder/guidance/6746fnl.pdf.

4. FDA/ICH (2006). Guidance for Industry. Q9 Quality Risk Management. Available at: http://www.fda.gov/cder/guidance/7153fnl.pdf.

5. Taguchi, G., Chowdhury, S., Taguchi, S. (2000). Robust Engineering, McGraw-Hill, New York.

6. Wu, Huiquan, Hussain, Ajaz S., and Khan, Mansoor A. (2007).Process Control Perspective for Process Analytical Technology: Integration of Chemical Engineering Practice into Semiconductor and Pharmaceutical Industries. *Chem. Eng. Comm.*, 194, 760-779.

7. Nair R, Vemuri M, Agrawala P, and Kim SI (2004). Investigation of Various Factors Affecting Encapsulation on the In-Cap Automatic Capsule –Filling Machine. AAPS PharmSciTech 5(4) Article 57. (http://www.aapspharmsciteh.org)

8. Wu, Huiquan and Hussain, Ajaz S. (2005). Integration of multivariate statistics and design of experiments to identify critical process variables for pharmaceutical process analytical technology (PAT) applications. In: American Statistical Association 2005 Proceedings of Joint Statistical Meetings, Minneapolis, Minnesota, August 7-11, 2005. CD-ROM , Mira Digital Publishing.

9. Carr RL (1965). Evaluating flow properties of solids. Chem. Eng. 72, 163-168.

10. FDA (2004). cGMP final report. Pharmaceutical cGMPS for the 21st Century—A Risk-Based Approach, Final Report. Available at : http://www.fda.gov/cder/gmp/gmp2004/CGMP%20report%20final04.pdf.

11. FDA Guidance (1995). Guidance for Industry. Immediate Release Solid Oral Dosage Forms. Scale-Up and Postapproval Changes: Chemistry, Manufacturing, and Controls, *In Vitro* Dissolution Testing, and *In Vivo* Bioequivalence Documentation. Available at: http://www.fda.gov/cder/Guidance/cmc5.pdf

AUTHOR INDEX

AUTHOR INDEX

AUTHOR INDEX